RFID-Enabled Sensor Design and Applications

Guest Editors

Piotr Jankowski-Mihułowicz
Mariusz Węglarski

MDPI

Basel • Beijing • Wuhan • Barcelona • Belgrade • Novi Sad • Cluj • Manchester

Guest Editors

Piotr Jankowski-Mihułowicz
Department of Electronic and Telecommunications Systems
Rzeszów University of Technology
Rzeszów
Poland

Mariusz Węglarski
Department of Electronic and Telecommunications Systems
Rzeszów University of Technology
Rzeszów
Poland

Editorial Office
MDPI AG
Grosspeteranlage 5
4052 Basel, Switzerland

This is a reprint of the Special Issue, published open access by the journal *Sensors* (ISSN 1424-8220), freely accessible at: https://www.mdpi.com/journal/sensors/special_issues/19P6506H43.

For citation purposes, cite each article independently as indicated on the article page online and as indicated below:

Lastname, A.A.; Lastname, B.B. Article Title. *Journal Name* **Year**, *Volume Number*, Page Range.

ISBN 978-3-7258-5963-4 (Hbk)
ISBN 978-3-7258-5964-1 (PDF)
https://doi.org/10.3390/books978-3-7258-5964-1

Contents

About the Editors

Piotr Jankowski-Mihułowicz

Piotr Jankowski-Mihułowicz (born 1974) received his M.Sc. in Electrical Engineering from Rzeszow University of Technology (RUT), Poland, in 1999, his Ph.D. from AGH University of Krakow in 2007, and his D.Sc. from the same institution in 2017. Since 1999, he has been working with the Department of Electronic and Telecommunications Systems, Faculty of Electrical and Computer Engineering, RUT. He is currently the Head of the Department of Electronic and Telecommunications Systems. His research focuses on radio frequency identification (RFID) systems, including their design, modeling, and practical applications. He collaborates with domestic and international companies to solve engineering challenges in RFID technology and related areas. Prof. Jankowski-Mihułowicz co-founded three specialized laboratories at RUT—RFID Technology, Electromagnetic Compatibility (EMC), and HYBRID Micro- and Nanotechnology—which form a comprehensive research base supporting joint academic–industrial projects. He is actively involved in technology transfer, industrial cooperation, and supervision of research and teaching initiatives. He co-organized the Polish Network of EMC Laboratories (EMC-LabNet), part of the national research infrastructure roadmap. Prof. Jankowski-Mihułowicz is the author of numerous high-impact publications and several patents, as well as a reviewer for international research projects and journals. He is a member of the IEEE (IEEE Antennas and Propagation Society), the International Microelectronics and Packaging Society (IMAPS), the Association of Polish Electrical Engineers (SEP), and the Polish Society of Theoretical and Applied Electrical Engineering (PTETiS). He has received multiple awards for scientific excellence and collaboration with industry, including the Podkarpacie Personality of the Year 2019 and 2022 (Science category) and the honorary title of Expert of the Polish Chief Technical Organization (NOT).

Mariusz Węglarski

Mariusz Węglarski was born in 1971. He received his M.Sc. degree from the Faculty of Electrical Engineering, Rzeszow University of Technology, Rzeszow, Poland, in 1996, his Ph.D. degree from the Faculty of Electrical and Computer Engineering, Rzeszow University of Technology, in 2005, and his D.Sc. degree from the Faculty of Electrical and Computer Engineering, Rzeszow University of Technology, in 2021. In 2021, he was an Associate Professor. Since 1996, he has been working with the Department of Electronic and Telecommunications Systems, Faculty of Electrical and Computer Engineering, Rzeszów University of Technology, as a Trainee Assistant from 1995 to 1996, an Assistant Professor from 1996 to 2005, an Assistant Professor from 2005 to 2019, and an Associate Professor since 2019. He is the author or co-author of over 100 scientific publications and four patents. He participated in over 30 research projects granted by the ministry, industry, and domestic and foreign institutions. His main research interests include hybrid electronics and microsystem technology, analysis of temperature fields, technology of RFID devices, and their practical applications. Prof. Węglarski is a member of the IEEE, the International Microelectronics and Packaging Society (IMAPS), the Association of Polish Electrical Engineers (SEP), and the Polish Society of Theoretical and Applied Electrical Engineering (PTETiS). He is a winner of many awards and honors for scientific activity and cooperation with the socioeconomic environment.

sensors

MDPI

Article

Active RFID Wake-Up Receiver Subsystem for Freight Wagon Localization Devices

Łukasz Krzak * and Cezary Worek

Institute of Electronics, Faculty of Computer Science, Electronics and Telecommunications, AGH University of Krakow, 30-059 Kraków, Poland; worek@agh.edu.pl

* Correspondence: lkrzak@agh.edu.pl

Abstract: This paper presents the concept, design, and performance analysis of an active radio wake-up and radio identification subsystem as part of an advanced localization device intended to operate within a large-scale freight wagon localization system. The system provides an efficient and cost-effective way to localize freight carriages, which, in the majority of cases, are currently not tracked. The localization device is battery-powered and uses an ultra-low-power radio interface for detecting wake-on-radio signals from nearby operator devices. The same interface is also used for communication within an ad-hoc wireless mesh network, which allows the localization devices to select the best device to send out localization information from the whole cluster through a cellular connection in order to minimize overall battery energy usage. The article presents the overall system architecture construction of the radio interface, including the wake-up subsystem, as well as the results of performance measurements.

Keywords: active RFID; wake-up receiver; WuRx; localization device; freight wagon positioning

1. Introduction

Railroad transportation is one of the most effective and environmentally friendly ways of transporting goods over long distances. According to the European Environment Agency, CO_2 emissions from rail transport are 3.5 times lower per tonne-kilometer than those from road transport [1]. However, the performance of rail freight transport in the EU remains unsatisfactory [2,3]. Among the many micro- and macroeconomic factors that affect this condition, one that is often described as challenging is the poor traffic management and suboptimal utilization of the current rail network and its assets [2]. One of the tools that helps improve logistic processes in this context is the automated and possibly real-time geolocalization of train carriages. However, in the case of freight wagons, such a system faces many technical challenges:

- It must operate at least over a continent, covering all possible train routes, coping with national requirements concerning radio spectrum usage;
- The localization devices must be mounted on each wagon individually, as the wagons are often swapped and mixed;
- The localization system should not rely on locomotive on-board units (LOBUs), as it complicates the adoption and deployment (more parties must be involved);
- Each localization device must be battery powered and the battery must last for at least 2 years (typical maintenance periods), as the freight wagons typically do not have access to a source of electric energy from the track [4].

As a result, many systems rely on specialized, battery-powered localization devices mounted directly on the wagons, utilizing satellite-based geolocation (GPS, Galileo, etc.) and cellular connection to send this information to the IT infrastructure. Some examples of such systems are SmartCargo from A1 Digital, deployed in Austria, and UDIV, deployed by the OLTIS Group in the Czech Republic. Much research and development has been conducted to build upon that architecture [5] or extend it [6].

Although using a cellular connection to upload wagon position to the cloud infrastructure has many benefits, such as worldwide coverage, high availability, and interoperability, it has one serious downside, which is relatively high energy usage. Based on an extensive one-year field study of such a prototype localization system, conducted by a major manufacturer of radiocommunication devices and systems for railways in Poland, Radionika, at least half of the energy available in the device was consumed by cellular connectivity. The main source of such high energy consumption is the long time (up to several minutes) required to establish the network connection. Data transmission uses a mature GPRS service, as the locations in which the train carriages operate are highly remote and at large distances from well-connected urban areas. The localization devices need to report their position at least once every 6 h and also when starting to move after a long stop. As a result, the batteries required for 2 year of operation become relatively large (two LSH 20 13Ah Li-ion batteries in the case of the above-mentioned prototype). This increases the cost of the device, makes the protective measures and maintenance harder, and significantly limits the miniaturization possibilities. Although some systems tried to use alternative uplink connection methods such as LoRaWAN [7] or are ready for modern long-range GSM services such as NB-IoT, the implementation of the required infrastructure in remote areas is progressing very slowly and does not yet provide adequate support for this kind of application.

Another requirement for the successful deployment of a freight wagon monitoring system, which is often overlooked, is the ability to easily identify and manage vehicles in the field. This requires the localization devices to be equipped with some sort of local communication interface that would allow the staff to interact with the device, which, in normal conditions, is mostly in deep energetic sleep mode, and thus is unable to receive any external signals. This brings even more benefits if the localization device is also combined with onboard sensors, which we will discuss in more detail in the next chapter.

The above-mentioned needs and challenges led to the collaboration between AGH University of Krakow and the Radionika company within an EU-cofounded project entitled "Research and deployment of a digital radio communication platform in railroad transportation". One of the key goals of the project was to design a localization system for railway rolling stock that could provide near real-time information about the current position and use of the wagons while addressing known challenges and limitations.

The developed solution is based on a standard approach in the following way:

- It uses battery-powered positioning devices mounted on each wagon;
- Each device uses satellite signals for geolocation and precise clock synchronization;
- Each device is equipped with a cellular module to communicate with the infrastructure.

However, in order to significantly improve energy efficiency, an additional low-power radio communication module was introduced, along with a mesh-based collaborative network protocol, allowing organizing the nearby devices into clusters and electing a single device from each cluster (based on a specific energy consumption metric) to send positioning information on behalf of all devices in the cluster. While it allowed reducing the number of cellular connections from each node almost 10 times, it also opened a way to solve the second problem, which is the lack of a method to establish local communication with the positioning device, that would be active all the time without consuming significant energy. Unfortunately, the minimum power required to run radio reception in standard monolithic integrated radio transceivers, operating in sub-GHz bands is in the order of tens of milliwatts (for example, the CC1125 (Texas Instruments, Dallas, TX, USA) radio transceiver used in the project requires 1–5 mA at 3.3 V to run in so-called sniff mode [8]), but the acceptable energy budget for the local ad-hoc communication requires methods consuming 3 orders of magnitude less, in the order of tens of microwatts. This is why the device was equipped with a custom, ultra-low power wake-up receiver, designed specifically for the stated application.

The rest of the paper presents in detail the resulting wake-up receiver subsystem and is organized as follows. In Section 2, we describe the designed freight wagon localization system focusing on the above-mentioned improvements and introduce the wake-up radio subsystem. In Section 3, we present a state-of-the-art method concerning wake-up radio detectors. In Section 4, we discuss the construction aspects of the radio subsystem. In Section 5, we present the results of measurements and tests and give our conclusions in Section 6.

2. Overview of the Freight Wagon Localization System

The solution developed by Radionika and AGH requires that each monitored wagon be equipped with a battery-powered localization device that periodically wakes up, reads its GPS coordinates, and sends this information via a cellular network to the dedicated IT infrastructure for further processing. However, the project method introduced some major improvements over existing solutions, which are depicted in Figure 1 and which we will briefly describe in the following sections.

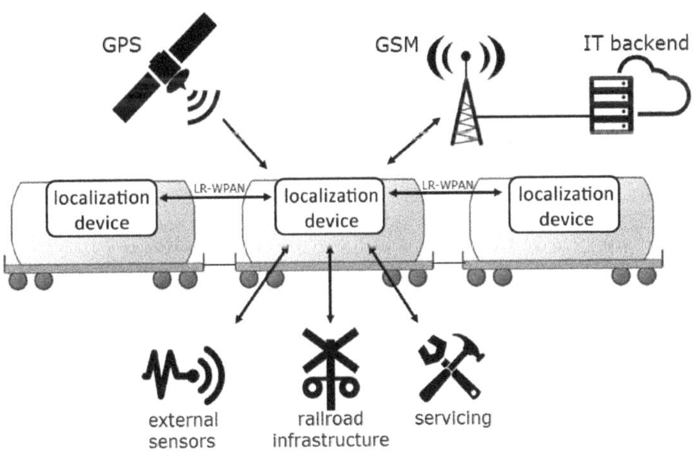

Figure 1. Illustration of the freight wagon localization system functionalities, featuring additional low rate wireless personal area network (LR-WPAN) radio interface.

2.1. Construction of the Localization Device

Figure 2 presents the block diagram of the localization device. The diagram was simplified in order to focus only on the components that are within the scope of this article. The components that constitute the wake-up radio subsystem and were designed specifically for this application are marked in green.

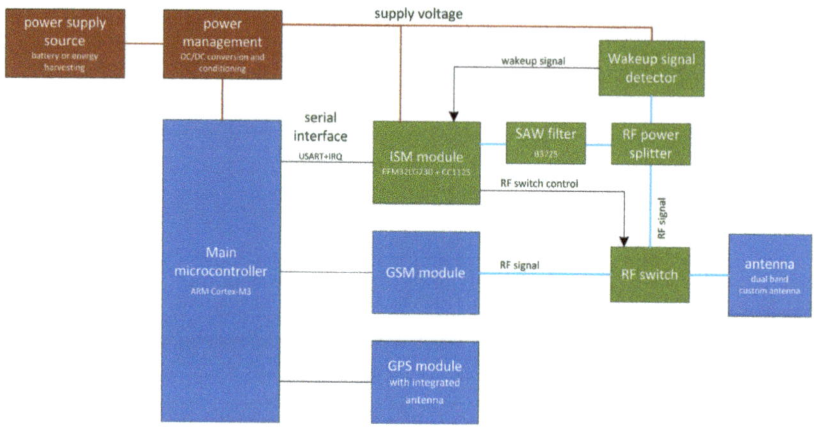

Figure 2. Block diagram of the localization device.

The localization device is powered with two LSH 20 13Ah Li-ion batteries (connected in series), but it could also be optionally powered by an energy harvesting module or a renewable energy source. The power management subsystem delivers operational voltages to all components. The system is controlled by an ARM Cortex-M3 main microcontroller. The platform integrates four different radio interfaces:

- GPS module with integrated antenna that receives satellite signal informing about location and time;
- GSM module that constitutes a long-range communication interface;
- Custom ISM (industrial, scientific, and medical band) module that delivers local communication capabilities between localization devices;
- Wake-up signal detector that enables peripheral sensors and other devices to asynchronously wake up the localization device and process external events.

It is worth underlining that the GSM and ISM modules, as well as the wake-up signal detector, are utilizing the same antenna circuit, which is a custom multi-band antenna. The antenna was designed to cover both the unlicensed 863–870 MHz frequency bands as well as the 900 MHz and 1800 MHz GSM frequency bands available in Europe. Figure 3 presents the VSWR of the antenna in the low bands.

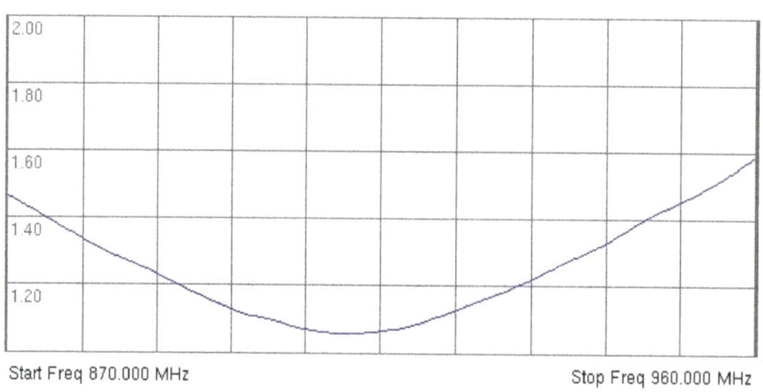

Figure 3. VSWR plot of the custom antenna designed for the localization device.

Figures 4 and 5 illustrate the construction of the localization device, including the enclosure and the default mounting method.

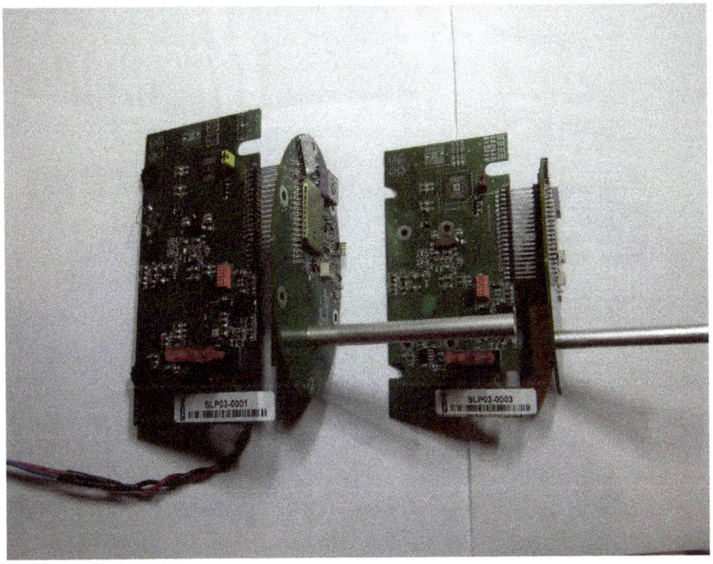

Figure 4. Pictures of the electronic part of the localization devices. The rounded top-mounted PCB includes the whole RF subsystem, including a custom antenna.

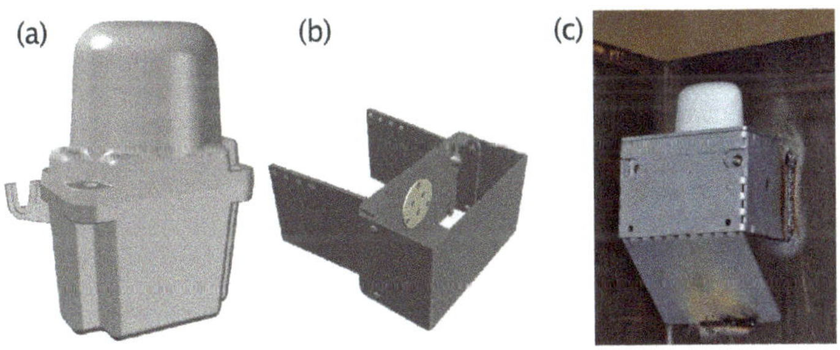

Figure 5. Pictures of the localization devices: (**a**) 3D model of the enclosure that holds the electronics, battery, and antenna, (**b**) 3D model of the steel pocket that is welded to the wagon side, (**c**) view of the localization device being mounted on a coal transportation wagon.

2.2. *Functionality Improvements over Existing Designs*

The introduction of an additional ISM radio interface coupled with a wake-up radio signal detector operating in an unlicensed 863–870 MHz band in the EU brought several new functionalities to the freight wagon localization system. Due to the fact that the train carriages usually operate in groups, adding a short-range wireless communication interface allows the establishment of a local, low-power ad-hoc communication network operating as an LR-WPAN (low-rate wireless personal area network). Such a network serves several purposes, which we will briefly discuss below.

2.2.1. Energy Savings Through Mesh Networking, Clustering, and Data Aggregation

A specialized, synchronous wireless communication protocol was developed, allowing the grouping of multiple nodes into clusters, aggregating data from such a cluster, and sending it via a cellular connection from a single cluster head node. The protocol has several interesting features:

- It allows devices (a.k.a. network nodes) to establish communication every fixed period of time (1 to 6 h). The ad-hoc operation session starts at predefined times in all localization devices thanks to GPS-based clock synchronization.
- It is stateless, meaning that each communication session is established ad-hoc, with no knowledge about past communication events. This allows the protocol to be insensitive to topology changes between sessions.
- The protocol allows the nodes to discover network neighbors and form the network within a fixed time (in the order of minutes).
- The protocol uses slotted TDMA with a fixed slot duration of 50 ms.
- The over-the-air baud rate is 100 kbps and the modulation characteristics follow the IEEE 802.15.4 SUN PHY specification.

The protocol organizes device operation into four phases:

- In the synchronization phase, the devices wake up roughly 3 min before the actual networking operation and use GPS receivers to obtain their position and synchronize the internal real-time clock to a global UTC time with approximately 5 ms accuracy. Next, they go to sleep until the exact moment the next phase is scheduled to start.
- In the advertising phase, which begins at nearly exactly the same moment in all nodes, the devices advertise their existence to other nodes, including information about the available charge. The goal of this phase is for each node to gather initial information about as many of its direct neighbors as possible.
- During the next phase called the association phase, the nodes use the information gathered during the advertising phase to form logical links over a slotted TDMA schedule. The goal is to assign to each pair of nodes a pair of time slots for exclusive communication, thus establishing an effective slotted communication schedule.
- In the final phase, called the clustering phase, the nodes use the established schedules to spread information about nodes within the possible cluster within a mesh network of a limited fixed radius so that collectively they can select the best candidate for the cluster head. Next, they send the request to this candidate to become the head of the cluster. If they receive confirmation, the clustering process ends. The cluster head is responsible for sending the position information about all the nodes that elected it to the infrastructure using a cellular connection.

The cluster head selection is completed based on a specific metric calculated by each node and broadcast during the clustering phase. This metric is based on the available charge in relation to the number of days the device operates since the last maintenance (when the battery is always replaced). This metric represents the deviation between the projected rate of battery charge depletion over time and the actual charge usage, recorded by the device. It is important to note that due to the fact that devices are rolled out in batches over time and the wagons are often swapped and mixed, many times the train will consist of devices in various stages of their projected lifetime. It is then crucial not to rely on the available charge metric, as the newest devices will always be depleted first. Instead, what we want to achieve is a uniform tempo of energy depletion, so devices with a slower charge depletion tempo are favored as cluster heads.

If, due to any circumstance, after the whole protocol operation is finished, the device fails to receive confirmation from the elected cluster head, it falls back to the default operation and sends its position through the GSM module on its own. However, through a detailed probabilistic modeling and behavioral simulation of the proposed protocol (out of the scope of this article), we have proven that this situation is very rare and the number of cellular transmissions from the group of nodes is vastly reduced. In the reference topologies studied, the charge required to relay the information about a wagon to the IT infrastructure, through the usage of clustering, is approximately 10 times smaller compared to the charge required for each node to send this information on its own, which is a significant energy savings [9].

2.2.2. Communication with On-Board Sensors

The presence of a short-range radio communication interface with a wake-up radio signal detector allows us to introduce external sensors mounted on the wagon that make it possible to sense the state of the carriage, as well as the load storage conditions, and send this information along with the localization data. These sensors can be powered from a battery [10] or even through power harvesting (e.g., using vibration energy) [11,12].

Within the described system, the sensors use asynchronous communication with the localization device. Whenever a sensor has data to send, it transmits a wake-up radio signal to activate the localization device, then sends the data packet and receives an acknowledgment. Failure to receive the acknowledgment forces the sensor to repeat the process up to two more times.

2.2.3. Communication with Railroad Infrastructure

Another function provided is the ability for the railroad infrastructure to request the localization device to report its position at the exact moment of passing a certain location, typically a mast with a stationary access point. This is useful from a logistics point of view whenever the arrival of a certain wagon at a predefined location must be recorded and time-stamped [13]. This operation of the access point is similar to a typical RFID interrogator [14]—it continuously broadcasts the wake-up radio signal (compliant with local duty cycle limitations) followed by an identification request and a listening period, in which the wagon localization device should respond. The response includes information about the wagon, including optional sensor data. The access point can now forward this information to the IT infrastructure using any wideband connection, for example, a cellular network.

2.2.4. Support for In-Field Service Operations

Due to reliability constraints and rigid enclosure, the localization device does not have a user interface (buttons, etc.) and is usually mounted in difficult-to-reach places on the train carriage (see Figure 5). As the localization devices are mostly in power-down mode and operate according to their own internal schedule, in order to connect to them on demand, e.g., for servicing purposes, the wake-up radio signal is used. The wake-up radio signal transmitter (including the radio module) is coupled with a mobile terminal. The service staff employee can trigger the transmission of the wake-up signal, which in turn wakes up the localization device mounted on a nearby wagon. The adjustment of output power and the design of the directional antenna can be used to limit the number of devices waking up in a train station environment, where many wagons are in close proximity. The session initiated through the service staff proceeds in almost the same way as the session with the access point described above.

Figure 6 illustrates all RFID features that are based on the wake-up radio subsystem.

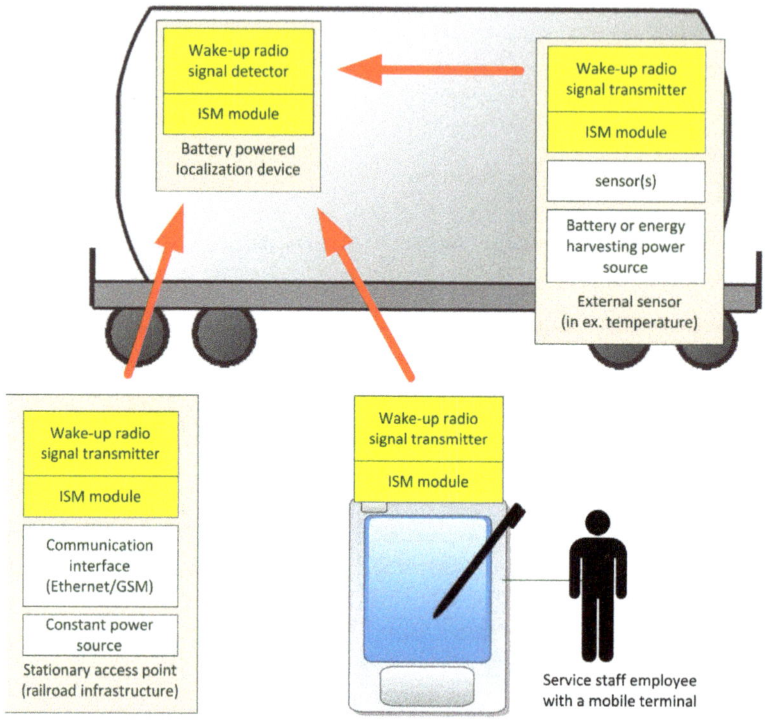

Figure 6. Illustration of the RFID features that are based on the wake-up radio subsystem. The red arrows show the wake-up signal path from various external devices, such as sensors on board of the wagon, stationary access points and handheld mobile terminals to the localization device.

3. Related Work

Wake-up radio (WuR) systems enable asynchronous wake-up of electronic devices, which are most often various types of autonomous sensors and nodes of wireless sensor networks (WSNs). Many WuR systems are based on an ultra-low power auxiliary radio path, most often called wake-up receivers (WuRx) [15–18]. Such receivers continuously monitor the RF spectrum for pre-specified event signatures (that is, wake-up messages) that tell the WuRx to enable the device to take further action, such as turning on the main radio or enabling a backscatter modulator [19]. In certain cases, WuRx systems allow for reducing the average energy consumption by continuous operation with low power consumption and activating the main broadband radio systems only after the occurrence/detection of an event and after transmitting the packets, switching back to the low-energy standby state.

Due to their practical application possibilities, they have been continuously developed for several decades, and the development of the Internet of Things (IoT) and Industrial IoT (IIoT) is currently expanding the areas of application [17,20,21]. The system solutions used in these designs are practically the same as in passive or semi-active RFID transponders; therefore, due to the diversity of applications, there are extensive articles in the literature presenting the current state of development of WuRx systems [15,17,20,22]. The practical dimension of the WuRx system solutions presented in specific applications, along with the description of the state-of-the-art method, can also be found in numerous doctoral theses [9,23–25].

In order to compare different WuRx solutions, many metrics need to be considered. The basic ones include the following [18,20]:

- Reading range: similar to radio systems, we can use the sensitivity metric, which is the easiest to compare. However, the effective communication range is greatly influenced by the selected operating band—most often used are unlicensed ISM bands: 125 kHz, 13.56 MHz, 433 MHz, 868/915 MHz, 2.4 GHz, and 5.8 GHz. Similar to RFID systems, the largest ranges are obtained in the 868/915 MHz band due to propagation conditions and legally permitted transmitter powers for the wake-up systems. For this reason, if we care about the reading range, comparisons should concern the same operating band and, additionally, the antenna system.
- Power Consumption: Since the WuRx system will be operating in standby mode all the time, it will often be a significant factor in the overall energy balance and, therefore, will heavily affect the battery lifetime of a device.
- Wake-up latency: The WuRx should respond to wake-ups with a reasonable delay, determined by the application requirements. In some applications, delays of a few seconds are acceptable, while in others, delays of microseconds are required. When designing the WuRx, there are ways to shape this parameter, but the general relationship is that the more stringent the latency requirements, the more power the WuRx receiver will draw.
- Immunity to interference and false alarms: WuRx systems may operate in an electromagnetic environment with high levels of interference, including many other radio emissions, and should, therefore, offer the ability to filter radio interference at an appropriate level.

In practical applications, it is also necessary to meet additional requirements related to the following:

- Resistance to environmental conditions: WuRx devices must operate reliably over a wide range of temperatures, humidity, and supply voltages, as well as in the presence of water, dust, and vibrations, and should not require any special calibration.
- Dimensions and weight: Minimizing these parameters in WuRx systems is strongly related to the economics and size of the antenna system and is especially important for small IoT/IIoT devices that rely on small batteries or energy harvesters.
- Economics: related to the time and costs of the project, the costs of starting production, the production itself, the availability of components, etc. In the simplest example, we can compare better technical parameters but more expensive solutions in the form of a specialized ASIC integrated circuit to solutions based on discrete elements [22].
- Compliance with standards: In order to introduce a product to the market, it is necessary to meet the legal requirements stated in relevant standards concerning electromagnetic compatibility, radio spectrum usage, environmental hazards, etc.

4. Construction of the Wake-Up Radio Subsystem

4.1. Selection of Operating Frequencies

Due to the principle of operation, where the wake-up radio signal activates the otherwise disabled localization device, causing it to broadcast a radio signal, the system needs to be treated as an RFID (radio frequency identification) system. Within EU countries, this poses restrictions and requirements that are stated in the ETSI 302 208 standard [26]. The standard divides the applicable 865–868 MHz frequency band into 15 channels, each 200 kHz wide. Channels 4, 7, 10, and 13 are high-power channels designated for RFID interrogators to broadcast the interrogator signal—see Figure 7. In these channels, up to +33 dBm (2W), effective radiated power is allowed for a maximum continuous time of 4 s, after which a minimum quiet period of 100 ms is required.

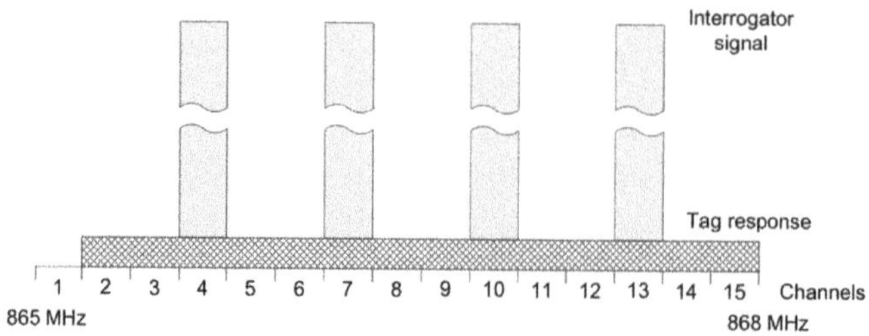

Figure 7. Channel plan for 865–868 MHz frequency band, according to ETSI EN 302 208 V3.3.1:2020.

In order to maximize the frequency separation between the wake-up signal and the communication signal, channel 4 was chosen. The wake-up signal is broadcast with a center frequency of 865.7 MHz, and the center frequency for the communication for the ISM module was selected to be 869.9 MHz.

4.2. Description of the Electronic Construction

The wake-up radio feature uses an active radio approach, in which a dedicated ultra-low-power radio signal detector is always on, listening to the radio channel and searching for a signal with certain expected attributes [15,22]. The presented patented wake-up radio subsystem is based on a dedicated radio signal detector, capable of listening while consuming very low energy. The architecture of the wake-up subsystem is presented in Figure 8, and detailed schematics are available in Appendix A.

The wake-up signal transmitter is a device that integrates a wake-up signal generator along with a reused ISM module and optional device-specific control logic (microcontroller) and/or sensors. In order to wake up a nearby localization device, the ISM module enables the wake-up signal generator that generates a special high-frequency signal (marked in red in Figure 8). The system uses a wake-up signal at 865.7 MHz with AM amplitude modulation (sinusoidal modulating signal, 217 Hz modulation frequency, 80% modulation depth). The band was chosen because of the possibility of using a single antenna for the ISM radio module, the GSM module, and the wake-up detector. The power levels indicated in red show the maximum wake-up signal power coming from the generator at various stages of the RF signal chain. The wake-up signal is fed to the antenna through a power splitter, which allows the signal to be isolated from the second signal path, which leads to the ISM module through an additional RF passband filter, providing even more attenuation on the wake-up signal and other possible strong interference signals (GSM, GSM-R). The green color indicates the power levels of the RF signal produced by the ISM radio module. Both signal levels were designed to meet the requirements concerning radio signal broadcast regulations, assuming the use of an antenna with up to 6 dBi gain.

Once the wake-up signal reaches the localization device (see Figure 8), it is detected by the wake-up signal detector. The signal from the antenna passes through a non-powered RF switch SKY13377-313LF (Skyworks, Irvine, CA, USA), which was selected due to its low attenuation, only 7.3 dB, in this mode of operation. The signal is then split into the ISM module through an additional B3725 SAW filter (TDK Corporation, Tokyo, Japan) as well as the actual detector circuit. At the detector input, the voltage of the signal is multiplied using DC-biased Schottky diodes (approximately 1.5 µA). As an envelope detector, a three-stage Dickson voltage multiplier was used in which Schottky diodes were polarized with a DC-biased current (approximately 1.5 µA) to increase their sensitivity [27,28]. The demodulated signal is then amplified by an ultra-low-power and low-noise discrete amplifier with series feedback, based on the BC847B and BC857B (NXP Semiconductors, Eindhoven, The Netherlands) transistor pair. Only on discrete elements has it been possible to build an amplifier with a very low current consumption of 2.4 µA and a sufficiently large gain bandwidth over 100 kHz. Next, it is filtered using a third-order Chebyshev low-pass filter and amplified using a pass-band amplifier (10× amplification, Q-factor = 5) based on OPA2369 (Texas Instruments, Dallas, TX, USA) with a current consumption of about 1.5 µA. The resulting signal is fed through two RC filters with different constant times to a comparator with hysteresis, which worked as a data slicer with a current consumption of about 0.8 µA, and then fed to the dedicated wake-up interrupt input of the ISM module. In case there is no wake-up signal present, the ISM module is put into deep sleep mode. Once the signal activates the comparator, the firmware running in the ISM module starts to count the frequency of the received impulses, as well as the number of periods, which further qualifies the signal as a valid wake-up. In the presented solution, the number of required valid impulses was set to 22, which gives the wake-up latency of approximately 106 ms. In case the signal matches all criteria, the ISM module is woken up to an active state and can further participate in the communication with the wake-up signal transmitter using the main mode of communication. In other cases, the module is put back into a deep, energetic sleep.

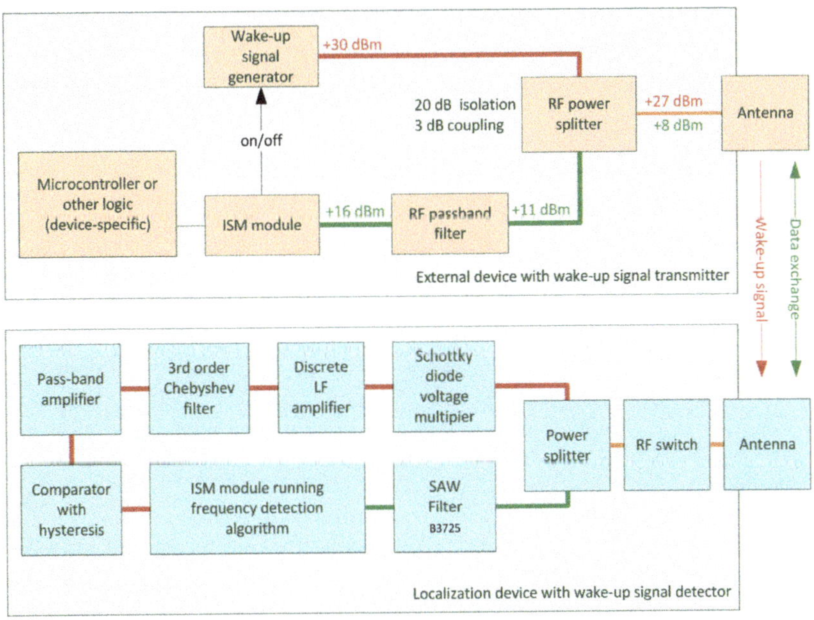

Figure 8. Illustration of the wake-up subsystem architecture. The top part presents an external device with a wake-up signal transmitter, capable of generating the wake-up signal (red and orange signal path) as well as transmitting and receiving data (green and orange signal path) once the localization device is woken up and the session is established. Red power values indicate power levels of the wake-up signal and green power values indicate power levels of the RF signal produced by the ISM radio module. An important aspect of the design is the RF power splitter, which has 20 dB isolation so that the wake-up signal leaking into the ISM module is at an acceptable level. The lower part of the diagram presents the components responsible for processing the received wake-up signal (red and orange signal path) and those that take part in regular data exchange (green and orange signal path) once the localization device is woken up and the session is established.

According to legal requirements, before emitting RF power, the wake-up signal transmitter checks if the radio channel is free using a listen-before-talk (LBT) algorithm. If it is, it broadcasts the wake-up signal for approximately 220 ms. This is sufficient, as the overall wake-up detection latency does not exceed 106 ms. Next, the device switches to reception mode and, after approximately 90 ms, broadcasts an IDENTIFY.request packet. This packet carries information about the wake-up transmitter address, as well as the length and number of time slots that immediately follow this packet, in which the woken-up localization devices may broadcast their responses. The length and number of slots are selected according to the application requirements. Figure 9 illustrates this process.

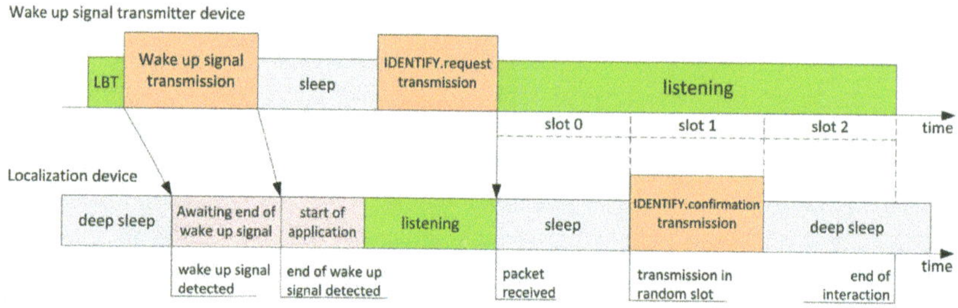

Figure 9. Illustration of the communication scheme between wake-up signal transmitter and woken up localization device. Green periods indicate state of radio listening, orange periods indicate state of radio transmission, grey periods indicate state of sleep and pink periods mark the activation of the wake-up signal detection algorithm.

When the wake-up signal is detected in the localization device, the algorithm waits for the end of the signal, and after some time needed to start the application, it switches the ISM module to reception mode. Once the IDENTIFY.request packet is received, the device chooses a random slot, in which it responds with the IDENTIFY.confirmation packet. This packet carries the address of the localization device. In this way, both devices exchange each other's identification data. Optionally, the IDENTIFY.request packet can carry a list of explicit node addresses that should maintain communication during the slotted channel access period. In such a case, only listed devices respond with an IDENTIFY.confirmation packet, which is useful in the particular case of a service device that needs to wake up only a specific localization device. The slotted channel access may continue to exchange additional application-specific information such as sensor values, etc. At this stage of project development, it is assumed that if the localization device being awakened receives important information from the sensors, railway infrastructure, or servicing device it is rather urgent and should be delivered to the IT infrastructure as soon as possible. This is why the localization device uses the GSM service immediately after the described interaction, to contact the supervising system and notify about the encountered event. In the future, it will also be possible to aggregate such data in the localization device and exchange them during the clustering process.

5. Measurement Results

5.1. Current Consumption

In the case of no wake-up signal, the detector was measured to consume no more than 7 µA of power supply current (when powered from 3 V). The current consumed by the MCU operating in shut-down mode is 5 µA (same supply voltage). Thus, the current consumed by the complete radio module in power-down mode with the active wake-up detector is approximately 12 µA (36 µW).

5.2. Sensitivity

For the purpose of measuring the sensitivity of the wake-up radio detector, a metric called the wake-up error rate (WER) was defined. Similarly to the packet error rate (PER), it is the ratio between failed wake-up radio transactions to the overall number of wake-up attempts. A failed wake-up transaction is the event in which the wake-up detector failed to recognize the wake-up signal, despite it being broadcast by the transmitter. Figure 10 presents the results collected from six prototype modules tested. For clarity, the measured points were linearly interpolated. As expected, the resulting plot resembles a typical characteristic of PER.

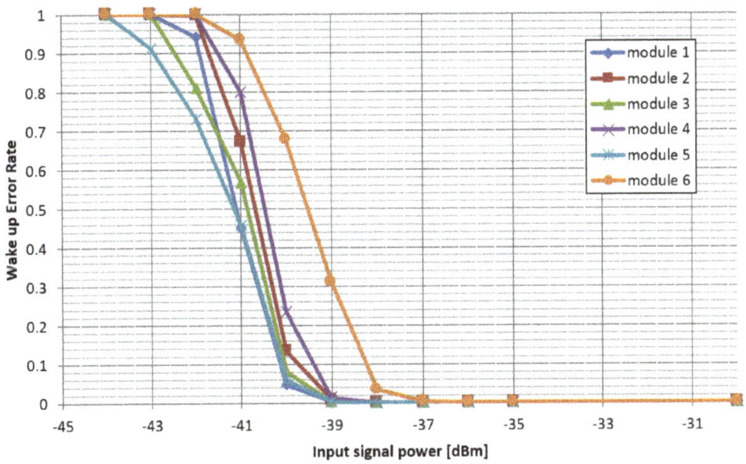

Figure 10. Results of sensitivity measurements.

The sensitivity of the WuRxs in this system has been evaluated to be −38 dBm, at which the ratio of reported missed or false wake-up events to the number of total wake-up events generated was below 1%. When evaluating this result against other wake-up radio solutions, we have to note that the effective sensitivity is reduced by the RF signal path that allows the use of a single antenna (see Figure 8), which attenuates the wake-up signal reaching the detector by 12–14 dB. When measured in a stand-alone configuration, WuRx has a sensitivity of −52 dBm. Taking into consideration that the solution consumes approximately 21 µW of power and is realized using discrete, common elements, the resulting sensitivity was considered satisfactory, especially since it shows high selectivity and resistance to interference signals.

5.3. *Effective Range*

In addition to the wake-up radio detector sensitivity measurements, a practical range test was also performed in three locations on the campus of AGH University in Krakow.

The tests were carried out under line-of-sight conditions in an open field location. In the test, a wake-up signal transmitter was sending 10 sessions of 1000 signal bursts each to the module that was operating as a wake-up detector. The distance between the transmitter and the receiver was changed, and the wake-up error rate was noted in the receiver. Figure 11 presents the results of a range test performed independently for six modules with a wake-up signal transmitter power of +27 dBm, AM modulation frequency of 217 Hz, modulation depth m = 80%, transmitter antenna gain +6 dBi, and receiver antenna gain +2 dBi. The wake-up error rate was averaged. The measured points were linearly interpolated for clarity.

In both cases, the practical range of the wake-up subsystem was estimated as 40–45 m, at which the WER metric is close to 1%.

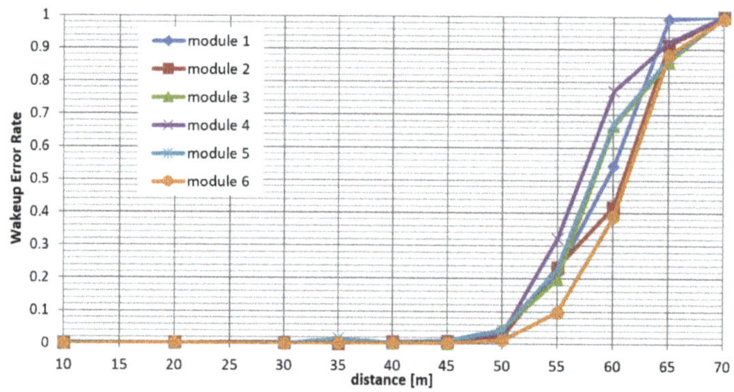

Figure 11. Results of range measurements.

5.4. Selectivity

In order to determine the frequency selectivity of the WuRx, a signal from an RF generator was fed to its input, with a level of −38 dBm. The signal was AM amplitude modulated, with a modulation depth of 80%. The modulating frequency was changed between 100 and 400 Hz, and the output signal was measured at the input of the U7A op-amp pin 1 (see Appendix A). This way, the signal path also included the diode detector. The results of the voltage amplification as a function of modulation frequency are shown in Figure 12.

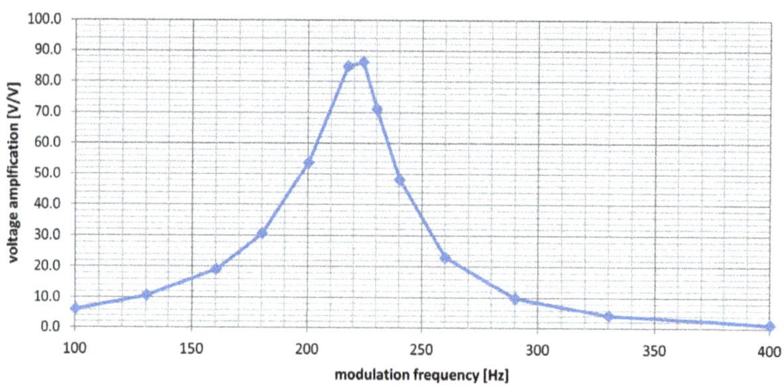

Figure 12. Results of selectivity measurements show the voltage amplification factor in the wake-up detector signal chain as a function of modulation frequency.

The presented frequency response was designed to provide strong rejection to interference signals that could falsely trigger the wake-up radio event. Frequencies above 300 Hz are heavily attenuated. The second part of the signal qualification, which also strengthens the selectivity, is done by the frequency detection algorithm running in the ISM module, as described in Section 4.2.

It is also worth noting that while the measurements presented in Figure 12 were taken at relatively low input power, the circuit shows good selectivity at input signal levels between −50 dBm and up to −20 dBm at which the wake-up detection happens only for modulation frequencies between 205 and 225 Hz. What is more, the circuit is not blocked with signals even up to 0 dBm input signal power, which is important in cases when auxiliary devices with wake-up signal transmitters are operating close to the positioning device.

6. Conclusions

This article presents a practical implementation of the WuRx low-power wake-up radio system that was successfully used in a railway wagon location system, and the presented solution architecture was unique enough to gain patent protection in Poland. It should be emphasized that the presented solution met all the requirements related to the operation of a radio device on the railway, i.e., the requirements of radio and railway standards, and functional tests were extended, among others, to include range tests in open space, where a distance of 45 m was achieved at 1% WER. The distinguishing feature of the presented solution is the operation of three independent radio systems (GSM, LR-WPAN network, WuRx) with a single antenna and an RF switch that is not powered and does not consume any current. However, this reduced the effective sensitivity of the WuRx system to a level of −38 dBm, even though it independently achieved a sensitivity of −52 dBm. The proposed design is based on the classic architecture with a voltage multiplier with pre-polarized Schottky diodes, and the low-frequency amplifier and filters were built from discrete elements. Due to the wake-up modulation frequency in the range of about 217 Hz and applied filtration, the selectivity of the wake-up system was ensured, and the wake-up latency parameter did not exceed 106 ms. The use of a sinusoidal modulation signal in the wake-up system transmitter caused the output signal to be spectrally narrow and pure, which in turn made the detection algorithms perform better. All tests performed confirmed the fully correct operation of the designed WuRx system, which did not require any tuning and added several improvements to the functionality of the entire freight wagon positioning system.

7. Patents

The localization device solution gained patent protection in Poland as "Method for radio communication and activation of a position locators and the position locator", Application number P.412566, Registration number Pat.228002, application date: 29 May 2015, registration date: 14 September 2017. Inventors: Cezary Worek and Łukasz Krzak, Applicant/Holder: Radionika sp. z o.o., Kraków, Poland.

Author Contributions: Conceptualization, C.W.; methodology and design, Ł.K. and C.W.; software, Ł.K.; validation, Ł.K. and C.W.; writing—original draft preparation, Ł.K. and C.W.; writing—review and editing, Ł.K.; visualization, Ł.K.; supervision, C.W.; project administration, Ł.K.; funding acquisition, C.W. All authors have read and agreed to the published version of the manuscript.

Funding: This research was funded by Operational Programme Innovative Economy in Malopolska region (Poland), grant number POIG.01.04.00-12-077/10-01.

Institutional Review Board Statement: Not applicable.

Informed Consent Statement: Not applicable.

Data Availability Statement: Data are contained within the article.

Acknowledgments: We would like to express our gratitude to the employees of Radionika company who supported our work and made this project possible, especially Andrzej Nikoniuk and Tomasz Rupiński.

Conflicts of Interest: The authors declare no conflicts of interest. The funders had no role in the design of the study; in the collection, analyses, or interpretation of data; in the writing of the manuscript; or in the decision to publish the results.

Appendix A

Detailed schematics of the wake-up signal detector.

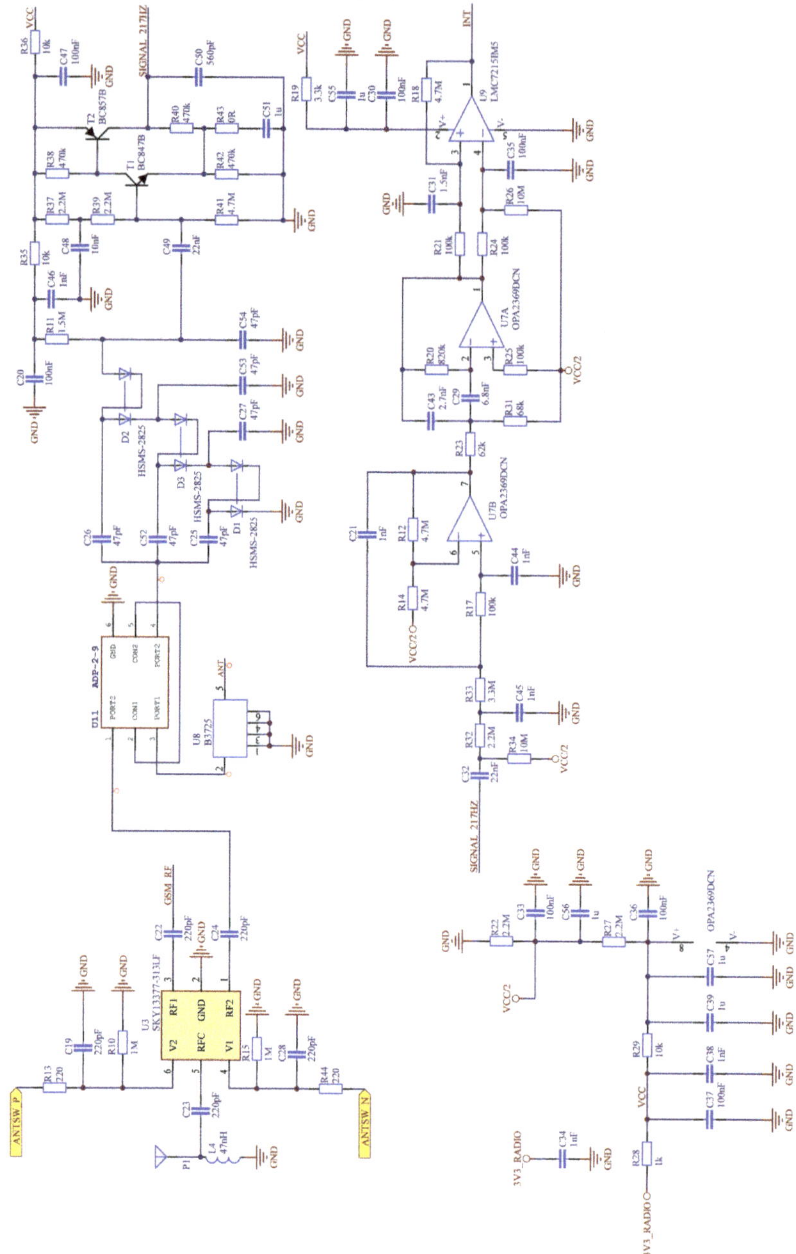

References

1. European Court of Auditors. *Rail Freight Transport in the EU—Still Not on the Right Track. Special Report No 08, 2016*; Publications Office: Luxembourg, 2016. [CrossRef]
2. The European Rail Research Advisory Council. *Research and Innovation—Advancing the European Railway, Future of Surface Transport Research Rail, Technology and Innovation Roadmaps*; Technical Report; The European Rail Research Advisory Council: Bruxelles, Belgium, 2016.
3. ECM Ventures. *European Rail Freight Market—Competitive Analysis and Recommendations: Study on Behalf of European Rail Freight Association (ERFA)*; Technical Report; European Rail Freight Association: Brussels, Belgium, 2022.
4. Lo Schiavo, A. Fully Autonomous Wireless Sensor Network for Freight Wagon Monitoring. *IEEE Sens. J.* **2016**, *16*, 9053–9063. [CrossRef]
5. Mazzino, N.; Brignole, O.; Cavalletti, C.; Terribile, S.; Diana, G.; Resta, F.; Tomasini, G. A geo-localization and diagnostic system for freight railway wagons. In Proceedings of the 11th World Congress on Railway Research, WCRR 2016, Milan, Italy, 29 May–2 June 2016.
6. Moya, I.; Perez, A.; Zabalegui, P.; de Miguel, G.; Losada, M.; Amengual, J.; Adin, I.; Mendizabal, J. Freight Wagon Digitalization for Condition Monitoring and Advanced Operation. *Sensors* **2023**, *23*, 7448. [CrossRef] [PubMed]
7. Paolini, G.; Fazzini, E.; Trovarello, S.; Amato, D.; Masotti, D.; Costanzo, A. An Innovative Multi-Port LoRa-Based Wireless Node for Railway Signaling and Positioning. *IEEE J. Radio Freq. Identif.* **2024**, *8*, 618–626. [CrossRef]
8. Johnsrud, S. *CC112x/CC120x RX Sniff Mode. Application Report SWRA428A*; Texas Instruments: Dallas, TX, USA, 2013.
9. Krzak, Ł. Energy Efficient Wireless Ad-Hoc Network for Applications in Railroad Transportation. Ph.D. Thesis, AGH University of Krakow, Kraków, Poland, 2018.
10. Bernal, E.; Spiryagin, M.; Cole, C. Onboard Condition Monitoring Sensors, Systems and Techniques for Freight Railway Vehicles: A Review. *IEEE Sens. J.* **2019**, *19*, 4–24. [CrossRef]
11. He, W.; Shi, W.; Le, J.; Li, H.; Ma, R. Geophone-Based Energy Harvesting Approach for Railway Wagon Monitoring Sensor with High Reliability and Simple Structure. *IEEE Access* **2020**, *8*, 35882–35891. [CrossRef]
12. Flankl, M.; Tüysüz, A.; Gong, C.; Stolz, T.; Kolar, J.W. Analysis and Modeling of Eddy-Current Couplings for Auxiliary Power Generation on a Freight Train Wagon. *IEEE Power Energy Technol. Syst. J.* **2018**, *5*, 139–147. [CrossRef]
13. Malakar, B.; Roy, B.K. Survey of RFID applications in railway industry. In Proceedings of the 2014 First International Conference on Automation, Control, Energy and Systems (ACES), Hooghly, India, 1–2 February 2014; pp. 1–6. [CrossRef]
14. Lee, L.T.; Tsang, K. An active RFID system for railway vehicle identification and positioning. In Proceedings of the 2008 International Conference on Railway Engineering—Challenges for Railway Transportation in Information Age, Hong Kong, China, 25–28 March 2008; pp. 1–4.
15. Fromm, R.; Kanoun, O.; Derbel, F. An Improved Wake-Up Receiver Based on the Optimization of Low-Frequency Pattern Matchers. *Sensors* **2023**, *23*, 8188. [CrossRef] [PubMed]
16. Weber, M.; Fersi, G.; Fromm, R.; Derbel, F. Wake-Up Receiver-Based Routing for Clustered Multihop Wireless Sensor Networks. *Sensors* **2022**, *22*, 3254. [CrossRef] [PubMed]
17. Zaraket, E.; Murad, N.M.; Yazdani, S.S.; Rajaoarisoa, L.; Ravelo, B. An overview on low energy wake-up radio technology: Active and passive circuits associated with MAC and routing protocols. *J. Netw. Comput. Appl.* **2021**, *190*, 103140. [CrossRef]
18. Piyare, R.; Murphy, A.L.; Kiraly, C.; Tosato, P.; Brunelli, D. Ultra Low Power Wake-Up Radios: A Hardware and Networking Survey. *IEEE Commun. Surv. Tutorials* **2017**, *19*, 2117–2157. [CrossRef]
19. Wang, P.H.P.; Zhang, C.; Yang, H.; Dunna, M.; Bharadia, D.; Mercier, P.P. A Low-Power Backscatter Modulation System Communicating Across Tens of Meters with Standards-Compliant Wi-Fi Transceivers. *IEEE J. Solid-State Circuits* **2020**, *55*, 2959–2969. [CrossRef]
20. Mercier, P.P.; Calhoun, B.H.; Wang, P.H.P.; Dissanayake, A.; Zhang, L.; Hall, D.A.; Bowers, S.M. Low-Power RF Wake-Up Receivers: Analysis, Tradeoffs, and Design. *IEEE Open J. Solid-State Circuits Soc.* **2022**, *2*, 144–164. [CrossRef]
21. Ketata, I.; Ouerghemmi, S.; Fakhfakh, A.; Derbel, F. Design and Implementation of Low Noise Amplifier Operating at 868 MHz for Duty Cycled Wake-Up Receiver Front-End. *Electronics* **2022**, *11*, 3235. [CrossRef]
22. Galante-Sempere, D.; Ramos-Valido, D.; Khemchandani, S.L.; del Pino, J. Area-Efficient Integrated Current-Reuse Feedback Amplifier for Wake-Up Receivers in Wireless Sensor Network Applications. *Sensors* **2022**, *22*, 1662. [CrossRef]
23. Mangal, V. Energy-Detecting Receivers for Wake-Up Radio Applications. Ph.D. Thesis, Columbia University, New York, NY, USA, 2020.
24. Sheshashayee, A.V. Wake-Up Radio-Enabled Wireless Networking: Measurements and Evaluation of Data Collection Techniques in Static and Mobile Scenarios. Ph.D. Thesis, Northeastern University, Boston, MA, USA, 2022.
25. Duvvuri, D. Ultra-Low-Power Wake-Up Receivers with Environmental Adaptive Designs Enabling Fully Integrated IoT Nodes. Ph.D. Thesis, University of Virginia, Charlottesville, VA, USA, 2023.

26. *EN 302-208 V3.3.1:2020-08;* Radio Frequency Identification Equipment Operating in the Band 865 MHz to 868 MHz with Power Levels up to 2 W and in the Band 915 MHz to 921 MHz with Power Levels up to 4 W. Harmonised Standard. European Telecommunications Standards Institute (ETSI): Sophia Antipolis, France, 2020.
27. Hewlett-Packard Co. *Schottky Diode Voltage Doubler, Application Note 956-4;* Hewlett-Packard Co.: Palo-Alto, CA, USA, 1995 .
28. Al-Azawy, M.M.; Sari, F. Analysis of Dickson Voltage Multiplier for RF Energy Harvesting. In Proceedings of the 2019 1st Global Power, Energy and Communication Conference (GPECOM), Nevsehir, Turkey, 12–15 June 2019; pp. 10–14. [CrossRef]

sensors

MDPI

Article

Effect of Embroidery Style on the Bandwidth of Textronic RFID UHF Transponder Antenna

Magdalena Nizioł [1,*], Piotr Jankowski-Mihułowicz [2,*] and Mariusz Węglarski [2]

[1] Department of Metrology and Diagnostic Systems, Rzeszów University of Technology, Wincentego Pola 2, 35-959 Rzeszow, Poland

[2] Department of Electronic and Telecommunications Systems, Rzeszów University of Technology, Wincentego Pola 2, 35-959 Rzeszow, Poland; wmar@prz.edu.pl

* Correspondence: m.niziol@prz.edu.pl (M.N.); pjanko@prz.edu.pl (P.J.-M.)

Abstract: The production of consumer electronics using electrically conductive materials is a dynamically developing sector of the economy. E-textiles (electronic textiles) are also used in radio frequency identification technology, mainly in the production of tag antennas. For economic reasons, it is important that the finished product is universal, although frequencies in radio systems have different values in different regions of the world. Therefore, the antenna bandwidth must be sufficiently wide so that the read range of the tag is maximally large for all frequencies of the specified band. The bandwidth of an antenna depends on its type and geometric dimensions, but this parameter can also be influenced by the way a given type of antenna is made. The authors prepared samples of embroidered RFID tag antennas for the UHF band using various types of embroidery. Then, its impedance and the read range of the tag were examined in order to determine the exact influence of the type of embroidery on the parameter of interest (antenna bandwidth). The results obtained during the research indicate the influence of different embroidery styles is present; however, that influence is not significant.

Keywords: RFID transponder; textronics; textile antenna; embroidery; embroidered antenna

1. Introduction

1.1. Purpose of the Work

Technological development is largely linked to changing consumer demands. Over the years, there has been an increasing emphasis on miniaturizing devices while maintaining a certain durability and, where possible, ensuring wireless operation. In order to meet such requirements, new fields of science have begun to develop, and one of them is textronics, which is a combination of, among others, electronics and textiles [1].

Nowadays, consumer electronics are increasingly integrated with electrically conductive textiles, which significantly expands the range of applications of individual devices [2–5]. One of the sectors of the economy in which e-textiles (combination of embedded electronics and textiles) are becoming increasingly important is the technology of radio frequency identification (RFID). Conductive materials (e.g., threads, fabrics) are used in the production of transponder antennas [6–11], thanks to which it is possible to integrate finished products with, e.g., items of clothing in a relatively easy and aesthetic manner. RFID tags with such antennas are widely used in other industries, such as health care [12–14] or the military [15,16].

A significant part of transponders is designed to operate in the UHF (ultra-high frequency) band [10,17–19]. Transponders for this band prevail over their counterparts in the low-frequency (LF) and high-frequency (HF) bands mainly in terms of the read range from about 10 m in the case of passive tags, to even 100 m in the case of active tags, with the appropriate selection of the antenna being of great importance.

The simplest type of antenna for UHF band RFID tags is a classic dipole antenna. By definition, however, it is a narrowband antenna, and in the UHF band, different frequencies are allocated to RFID devices depending on the region of the world, e.g., for the region covering Europe (region 1, according to frequency allocation by International Telecommunication Union), it is 862–870 MHz and for Asia (region 3), it is 950–956 MHz [20]. Due to this frequency dispersion, the use of narrowband antennas is uneconomical, as a tag designed to operate in one of the previously listed regions may be incompatible with RFID systems in another region, which may necessitate the production of additional solutions.

From an economic point of view, the ideal situation is to develop a solution with a wideband antenna so that the tag works effectively regardless of location. An example of such an antenna can be the bow-tie antenna, which is still an object of interest for engineers and scientists [21–28].

A bow-tie antenna is a planar antenna. Making it in the form of a textile antenna requires embroidering a relatively large surface of the radiator, which translates into a significant amount of material used (e.g., conductive thread). In mass production, this can generate considerable costs, which can be reduced by optimizing the antenna embroidery process.

The main parameter on which the authors focus their attention is the bandwidth of the antenna, so that after integrating it with the chip, the finished tag can effectively work in the frequency band from 860 to 960 MHz (UHF band coverage for all regions of the world). The authors are looking for a relationship between this parameter and the style of embroidery used to produce the model antennas. It was hypothesized that a change in the style and density of embroidery would cause a change in bandwidth, but a properly optimized embroidery process could reduce the amount of material used without significantly deteriorating the parameters of the tested antenna.

Similar studies have been found in the literature, but they either do not directly concern antennas [29], or focus on the relationship between the type of embroidery and other parameters, e.g., the shape of the antenna [30]; tag read range [31–35]; impedance [34]; gain, efficiency and radiation pattern [36–39]; reflection coefficient [36–38,40] (indirectly related, but not exhaustive of the subject under study).

1.2. Textile UHF RFID Transponder Antenna

The research team of the co-authors of the article has been dealing with the subject of RFID for many years, and the result of their research is, among others, a patent (patent no PL 231291 B1 Polish Patent Office) concerning a textronic tag (Figure 1) [41]. According to the idea behind this solution, the antenna and chip are not galvanically connected. Instead of a physical connection carried out, e.g., by means of a conductive thread from which the tag antenna is sewn, two coupling circuits in the form of loops are used.

A solution of a textronic tag for the UHF band with a dipole antenna was developed and tested in many aspects [42–44]. As part of this article, the dipole antenna is replaced by a wideband bow-tie antenna. In the design, the length of the antenna corresponds to the length of the classic dipole being replaced. An increasing bandwidth is ensured by the expansion of the radiator. The antenna is connected to the chip in a manner analogous to the previous solution, i.e., by means of a loop with a diameter of 5.5 mm ensuring inductive coupling with the microelectronic circuit.

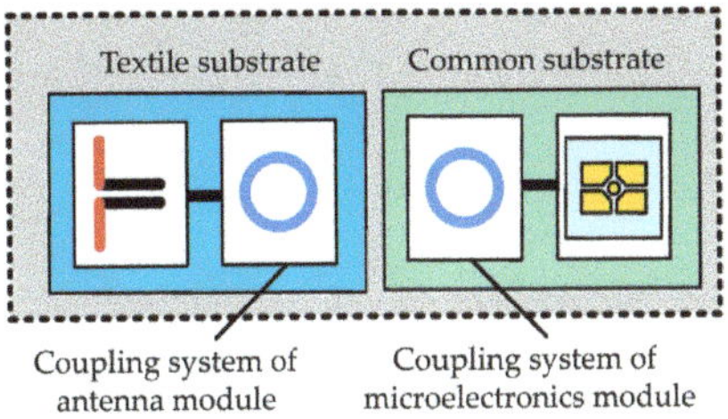

Figure 1. Block diagram of textronic RFID transponder (RFIDtex tag).

Due to the fragment of the hypothesis presented in Section 1.1 about the limitation of the use of material to create an antenna, in the experiments, in addition to the planar model, contour models were also taken into account. The geometry of the developed models is shown in Figure 2.

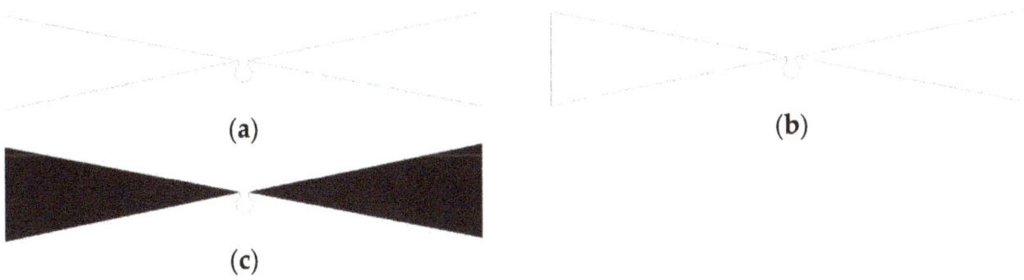

Figure 2. Geometry of designed antennas: (**a**) open contour; (**b**) closed contour; (**c**) planar.

Contour models were prepared for the analysis, including open (a) and closed (b), as well as a planar model (c), under which, by means of the style and density of embroidery, the degree of filling of the radiator will realistically be changed.

To ensure the proper operation of the RFID tag, a proper impedance match between the antenna and the chip is necessary. Both the antenna impedance (Equation (1)) and the impedance of the chip (Equation (2)) are expressed in complex form and are conjugates of each other. The quality of this conjugation is expressed numerically by power transfer coefficient τ (Equation (3)).

$$Z_{TA} = R_{TA} + jX_{TA}, \tag{1}$$

$$Z_{TC} = R_{TC} + jX_{TC}, \tag{2}$$

$$\tau = \frac{4Re(Z_{TA})Re(Z_{TC})}{(Re(Z_{TA} + Z_{TC}))^2 + (Im(Z_{TA} + Z_{TC}))^2}, \tag{3}$$

In Equations (1) and (2), the symbols R_{TA} and R_{TC} stand for antenna resistance and chip resistance, respectively, while X_{TA} and X_{TC} represent reactance. Symbols Z_{TA} and Z_{TC} stand for antenna and chip impedance.

2. Materials and Methods

In order to create samples of the analyzed antenna, its geometry was first implemented into the PE DESIGN 11 program, which is dedicated to Brother (Brother Industries, Ltd., Nagoya, Japan) embroidery machines. In this software, in addition to designing the pattern and configuring key parameters of the embroidery, there is also an option to simulate the embroidery process. The view of the first models prepared for embroidery is shown in Figure 3.

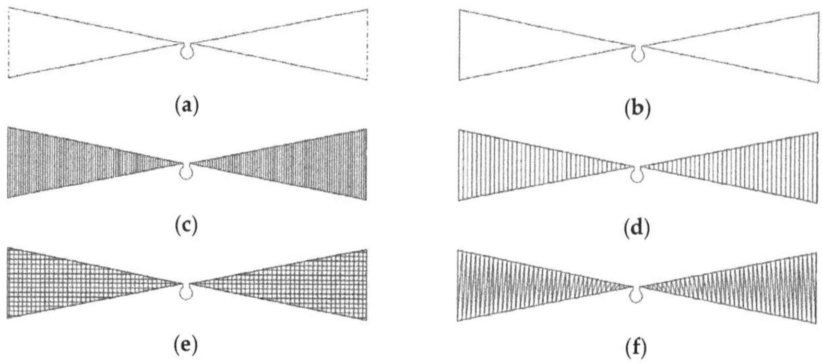

Figure 3. Models prepared in PE DESIGN 11: (**a**) model MN1—open contour; (**b**) model MN2—closed contour; (**c**) model MN3—planar ver. 1; (**d**) model MN4—planar ver. 2; (**e**) model MN5—planar ver. 3; (**f**) model MN6—planar ver. 6.

The first two samples show just the outline of the antenna. The next 4 are a filled model, but the samples differ in the style of stitch filling the surface of the radiators and the density of the stitch. The exact sewing parameters of individual models are presented in Table 1.

Table 1. Detailed information about used styles of embroidery.

Model	Stitch Style	Density	Stitching Direction
MN1	Running Stitch	2 mm	-
MN2	Running Stitch	2 mm	-
MN3	Fill Stitch	1.0 line/mm	90 deg
MN4	Fill Stitch	0.5 line/mm	90 deg
MN5	Net Fill Stitch	2 mm	0 deg
MN6	Satin Stitch	1 line/mm	90 deg

The MN3 pattern most closely represents the full filling of the radiator surface. Each successive pattern is realized with a lower density, resulting in less use of the conductive thread. However, in the case of patterns with vertical seams, there is a risk of too weak an electrical connection between successive seams, so the MN5 pattern was implemented in the form of a grid; horizontal stitching should significantly improve the electrical connections between successive sections of thread.

The designs shown in Figure 3 were embroidered using the BROTHER INNOV-IS V3 (Brother Industries, Ltd., Nagoya, Japan) embroidery machine. The conductive thread was placed in the bobbin (bottom thread), while the upper thread was a non-conductive polyester sewing thread. All antennas were embroidered at the speed of bobbin 350 rpm, with the tension of the thread set at level 9. Conductive threads were used to create the samples, the basic parameters of which are presented in Table 2. All patterns are embroidered on linen fabric. The appearance of the finished samples is shown in Figure 4.

The threads (the ones marked as A and B) used, in addition to parameters presented in Table 2, also differ in stiffness, which at the production stage can affect the final effect of the embroidery process—the stiffer the thread, the more difficult it is to arrange it into more complex patterns. In addition, the number of needle insertions should be properly selected—too small a number may cause deformation of the pattern, which was the case with the MN6 pattern (Figure 5).

Table 2. Conductive thread parameters.

Symbol	Manufacturer	Brand Name	Material	Thread Resistivity, Ω/m
A	Syscom Advanced Materials, Inc., Columbus, OH, USA	Liberator 40	Silver coated vectran	3.28
B	Plug&Wear, Florence, Italy	PW018A	Silver plated nylon	300
C	Sparkfun Electronics, Niwot, CO, USA	DEV-11791	Stainless steel	91.86
D	AMANN Group, Bönnigheim, Germany	Silver-tech 30 tex 96	Silver coated polyamide/ polyester	85

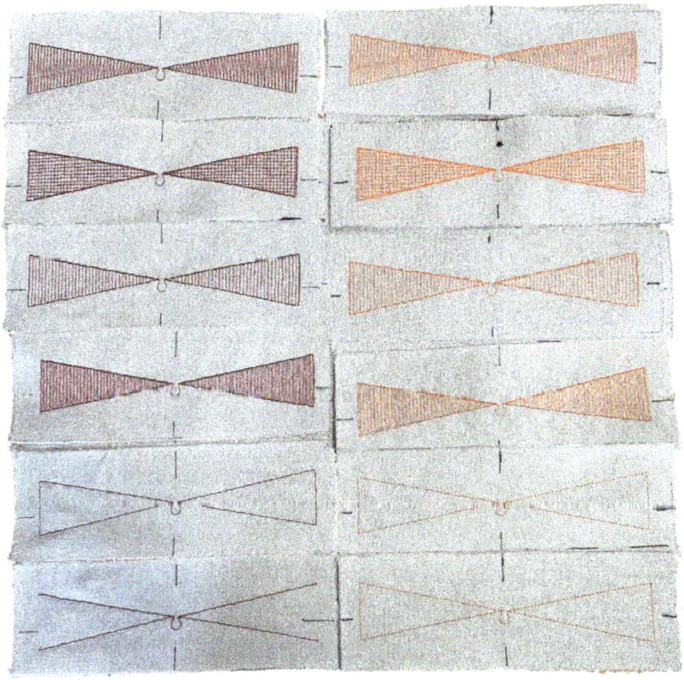

Figure 4. Photograph of sewn samples (red ones—thread B; orange—thread A).

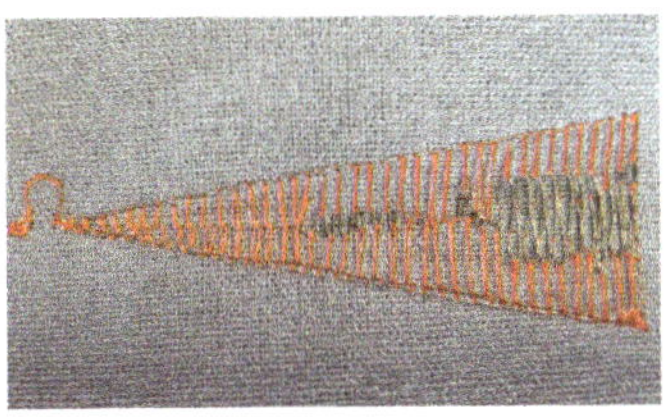

Figure 5. Photograph of a deformed sample (model MN6).

Although seemingly the pattern can be considered completely unsuccessful and the sample useless, the antenna was eventually used for research, because part of the radiator is embroidered (patch at the edge) and it has an electrical connection to the contour and coupling circuit.

To make the research more comprehensive, a decision was made to extend the experiments with additional formulas (Figure 6) and threads (marked as C and D in Table 2). Additional samples were sewn with threads with C and D markings on the same substrate material.

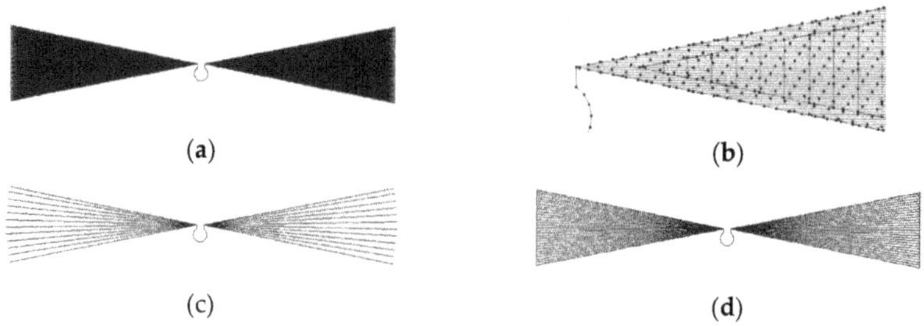

Figure 6. Models prepared in PE DESIGN 11: (**a**) model MN8—planar ver. 5; (**b**) model MN8—closeup; (**c**) model MN9—planar ver. 6; (**d**) model MN10—planar ver. 7.

Finally, a total of 42 samples divided into 4 groups (depending on the conductive thread used) were prepared for testing. However, the main object of interest was the groups A and B containing MN1–MN6 models. Groups C and D (the appearance of selected samples is shown in Figure 7) complementing the analysis contain MN1, MN4 and MN5 models from the basic pool of projects and additionally, MN8, MN9 and MN10 models from the additional pool (group D only). At this point, it should also be mentioned that group C was used only for some of the experiments.

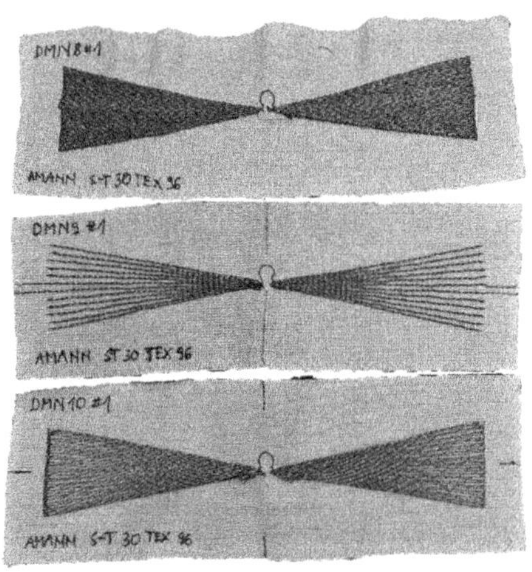

Figure 7. Photograph of sewn samples (selected models from group D).

The prepared samples were used for two stages of research—first, the impedance of the model antennas was measured, and then the read range of the tag with the model antenna was examined. The appearance of the measuring stations used during the tests is shown in Figure 8.

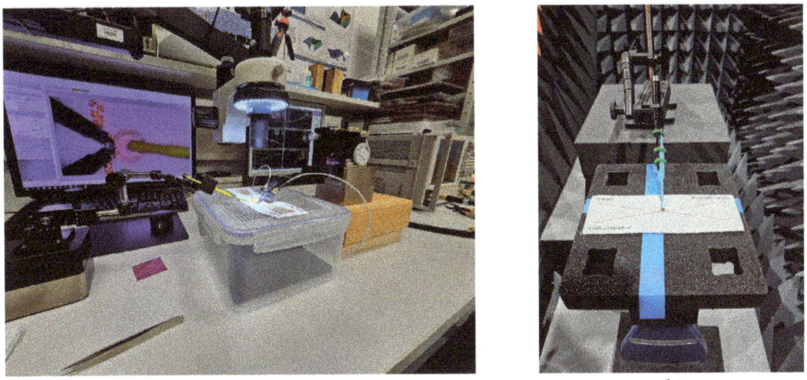

(a) (b)

Figure 8. Photographs of the measurement stands: (**a**) impedance measurements; (**b**) anechoic chamber.

For impedance measurements (Figure 8a), a Keysight (Keysight Technologies, Inc., Santa Rosa, CA, USA) PNA-X N5242A vector network analyzer with attached PacketMicro DPSS201505 SS05–0053 probes was used. A microscope was additionally used to check the correct attachment of the probe to the measuring circuit. The read range measurements were carried out in the Microwave Vision Group (Paris, France) anechoic chamber equipped with the Voyantic (Voyantic Ltd., Helsinki, Finland) Tagformance Pro measurement system. This system is a comprehensive combination of hardware and software (Tagformance Pro UHF, currently in version 14.0) dedicated to RFID tag measurement. Measurements were made in accordance with the ISO IEC 18000-63 standard. The set parameters of the reader are shown in Figure 9.

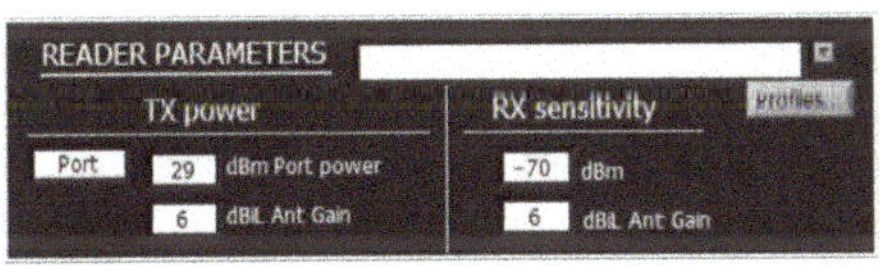

Figure 9. Reader parameters in Tagformance.

3. Results

The aim of the study was to determine whether the style of embroidery used to produce a planar antenna affects its parameters, in particular the bandwidth. Changes in the bandwidth value can directly affect the performance of the tag under certain conditions.

In order to determine the bandwidth, the impedance of the model antenna must first be measured. The impedance measurements of individual samples were made in the frequency range of 0.5–1.2 GHz. The results obtained are presented in Figure 10.

Between the individual models, the change in the antenna resonance frequency is visible (dispersion between about 550 and 700 MHz). The values for planar models are similar, but for contour models, and in particular for open contour models, the value is much higher. This is mainly related to changes in the R, L and C parameters of the antenna. The impedance value also changes, with the trend of changes being analogous. An important observation during the analysis is the significant similarity in the shape of the waveform and the measured values in the case of planar models, which may suggest that there is no need to fully embroider the radiator to achieve similar antenna parameters.

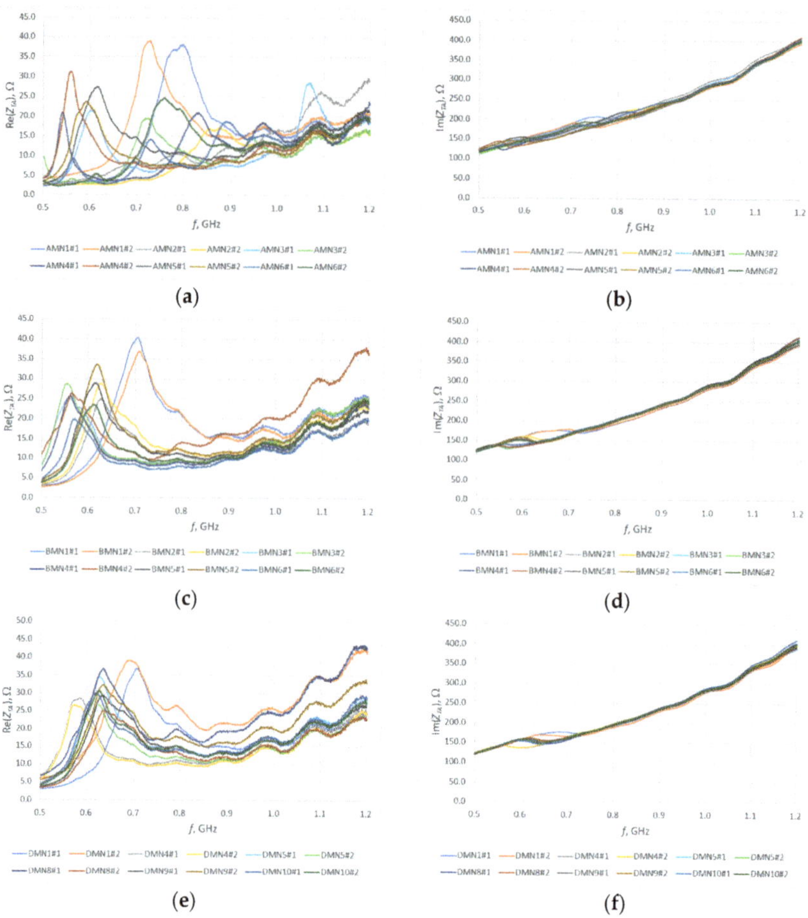

Figure 10. Impedance of the model antennas: (**a**) group A, real part; (**b**) group A, imaginary part; (**c**) group B, real part; (**d**) group B, imaginary part; (**e**) group D, real part; (**f**) group D, imaginary part.

The observed differences between samples of the same model result directly from the production process of the model antenna. During sewing, not only the thread from which the antenna is sewn, but also the substrate material is subjected to tension. Each deformation causes some dispersion of antenna parameter values—the largest differences are visible in the case of group A samples; in other cases, the samples are very similar to each other and the differences are negligibly small.

A good parameter from which to determine the bandwidth of an antenna can be the power transfer coefficient, often also referred to as conjugate match factor (CMF) (Equation (3)). This coefficient describes the quality of the antenna's fit to the chip, and its value ranges from 0 (no match) to 1 (perfect match). The impedance of the chip (Figure 11)—SL3S1214 UCODE 7m in case SOT886 manufactured by NXP Semiconductors, Eindhoven, The Netherlands—necessary to determine the CMF was previously determined by numerical calculations (linear regression on data from chip datasheet). The selected chip operates in passive (no additional power supply) mode and exactly this solution is tested. Figure 12 shows the CMF values calculated for the model antennas as a function of frequency.

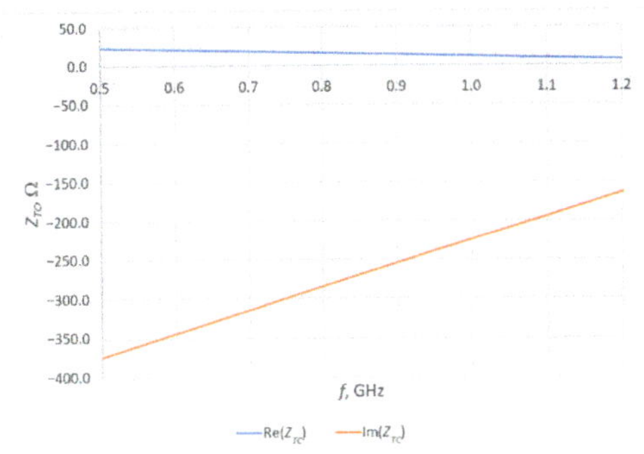

Figure 11. SL3S1214 UCODE 7m chip impedance.

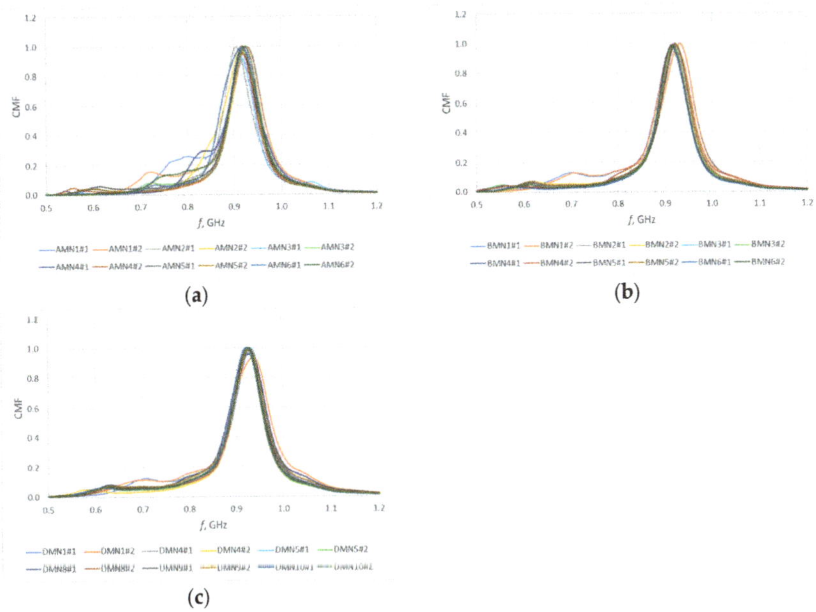

Figure 12. Calculated CMF: (**a**) calculations for group A; (**b**) calculations for group B; (**c**) calculations for group D.

It should be noted that the impedance of the chip changes with frequency. In addition, the imaginary part takes on relatively large values, which consequently slightly increases the challenge of perfect antenna matching.

The analysis of the presented courses shows a large convergence of the obtained results. An important observation, nevertheless, is the increase in resonant frequency relative to the values observed in the antenna impedance plots. This is due to the fact that the presented results are for a resonant circuit consisting of an antenna and a chip. However, from the point of view of the purpose of this publication, it is important whether the bandwidth is changed, not the CMF that was used to determine it. This bandwidth B (Figure 13) is determined as the difference between the upper (f_H) and lower (f_L) frequency at which the parameter (in this case CMF) reaches 70% of its maximum value (within the assumed range of best fit, the antenna will provide the most effective operation of the tag). The results obtained are presented in a graphic form in Figure 14.

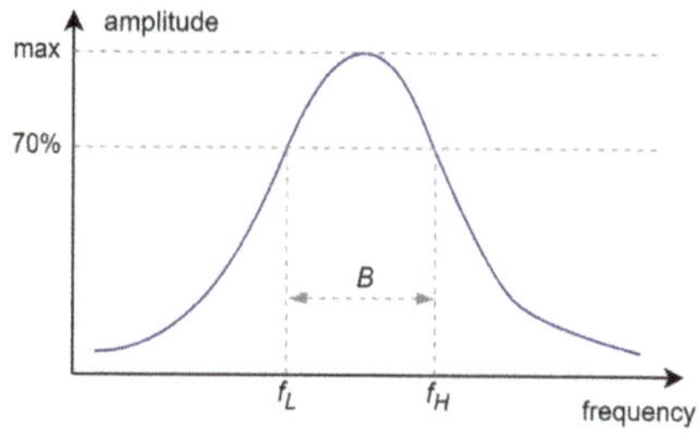

Figure 13. Methodology of determining model antenna bandwidth.

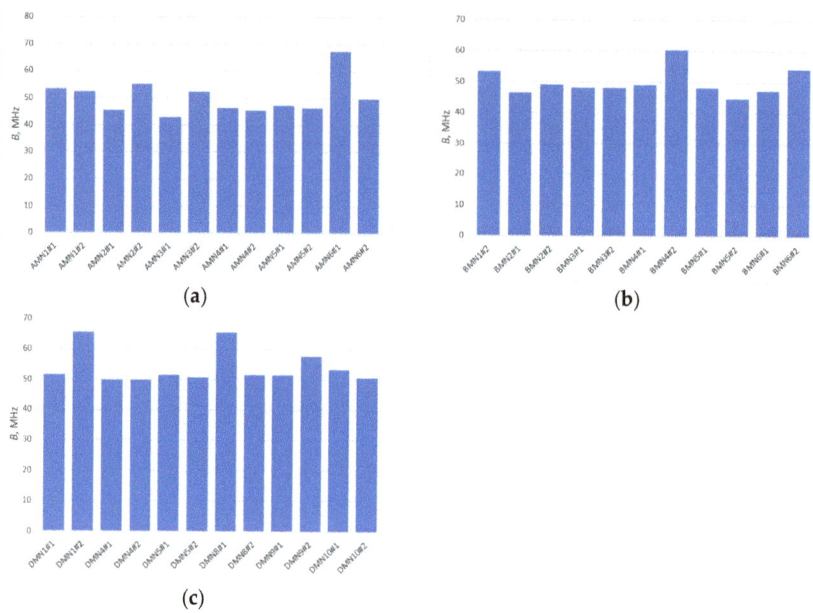

Figure 14. Graphical representation of model antenna bandwidth (determined from CMF) values for: (**a**) samples from group A; (**b**) samples from group B; (**c**) samples from group D.

The results obtained are characterized by a certain spread of values. By determining the standard deviation (Equation (4)) and then the relative standard deviation (Equation (5)) from the obtained results (Table 3), a spread around the mean in the range from 9% (group B) to 13% (group A) is obtained. It is worth noting that in the results obtained, however, there are single values that differ significantly from the others. Comparison of the questioned values with the values for the second samples of a given model shows a relatively large discrepancy; therefore, these values can be treated as parasitic errors caused by, for example, deterioration of the coupling between the antenna and the chip coupling system during the measurement (shifted chip substrate). Eliminating these values from the analysis will cause the spread of bandwidth values of the model antennas to be even smaller, and this leads to the conclusion that the type of embroidery has a limited impact on the antenna's bandwidth.

$$s = \sqrt{\frac{\sum(B_i - B_{av})^2}{n-1}}, \tag{4}$$

$$s_{rel} = \frac{s \cdot 100}{B_{av}}, \tag{5}$$

where B_i stands for individual sample bandwidth, B_{av} is the average bandwidth for all samples within group and n is the number of samples in the group.

Table 3. Standard deviation and relative standard deviation calculated for bandwidth determined from CMF.

Group of Samples	s, MHz	s_{rel}
A	6.57	13.05%
B	4.46	8.95%
D	5.74	10.59%

The model antennas were prepared for the UHF RFID tag, so the last stage of the research was to measure the read range of such a tag in the anechoic chamber. The waveforms recorded in the UHF Tagformance software (for selected samples) are shown in Figures 15–17.

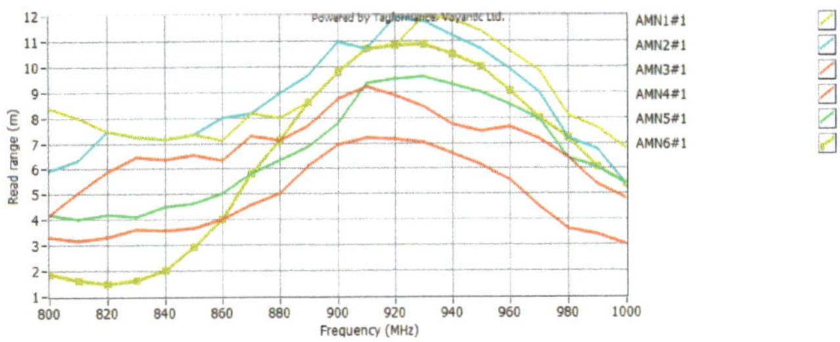

Figure 15. Read range of UHF RFID tag with model antenna—data obtained for group A.

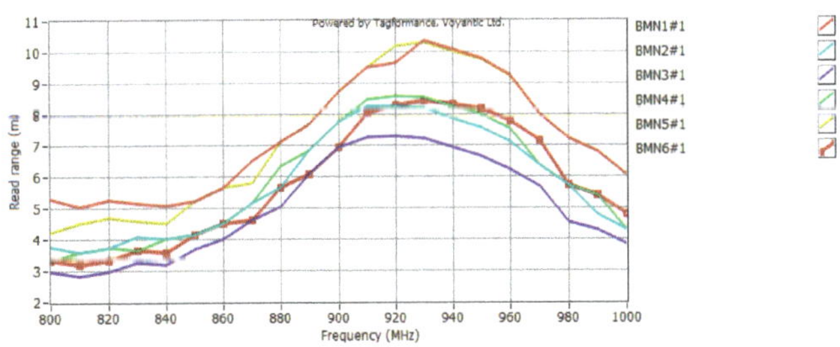

Figure 16. Read range of UHF RFID tag with model antenna—data obtained for group B.

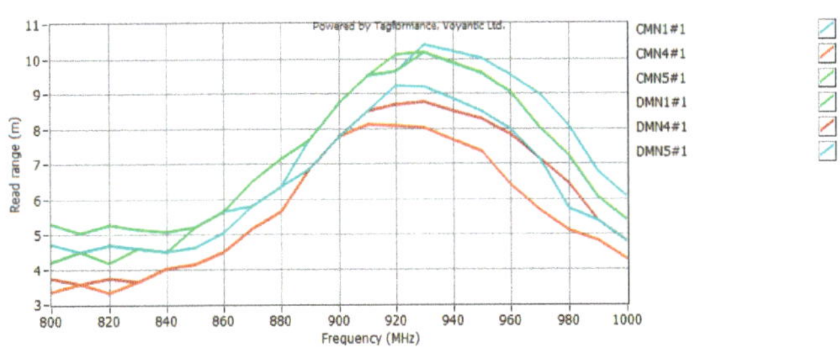

Figure 17. Read range of UHF RFID tag with model antenna—data obtained for groups C and D.

Analyzing the obtained plots, it can be seen that the maximum read range is shifted towards the upper limit of the UHF band. This could be expected from an analysis of the CMF graphs, since the frequency for which the antenna is best matched (CMF equal to or close to 1) is also shifted toward the upper limit of the band. However, the bandwidth guaranteed by the bow-tie antenna ensures the operation of the tested tag in the entire analyzed band (860–960 MHz). The analysis of the obtained plots in terms of bandwidth (B_{RR}) values of the tested antennas (carried out analogously to the analysis on the basis of CMF) (Figure 18) confirms the previous conclusions. The difference between the patterns is not very significant, i.e., a relative standard deviation (calculation results shown in Table 4) between 4% for part of group D (only three models from the basic pattern pool were examined) and 12% for group A. Therefore, there is no need to fully embroider the designed geometry. The type of embroidery can be selected so as to minimize the amount of material used, and thus reduce the cost of the antenna production.

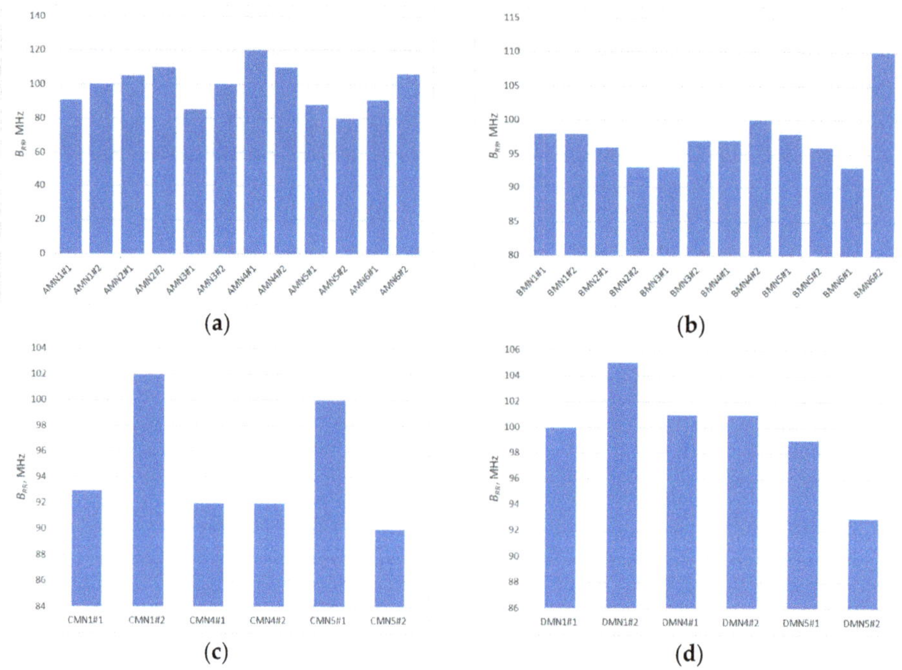

Figure 18. Graphical representation of model antenna bandwidth (determined from read range) values for (**a**) samples from group A; (**b**) samples from group B; (**c**) samples from group C; (**d**) samples from group D.

Table 4. Standard deviation and relative standard deviation calculated for bandwidth determined from read range.

Group of Samples	s, MHz	s_{rel}
A	11.97	12.11%
B	4.56	4.68%
C	4.92	5.18%
D	3.92	3.93%

While the results obtained are satisfactory, it is possible to significantly improve them (especially in the context of the bandwidth) thanks to better impedance matching of the model antennas. Then, it would be possible to bring the curve to a complete flattening in the entire band.

4. Discussion

The aim of the study was to investigate the effect of the style of embroidery used to create textile planar antennas on the bandwidth. The motivation for undertaking the research was to determine the possibility of optimizing the economic aspect of the antenna production process. The cost of conductive threads, especially for mass production, is significant, so the amount of material used is extremely important.

In order to meet the assumed goal, several versions of the bow-tie antenna were first prepared, including an antenna model in the form of an open contour (MN1), a closed contour (MN2) and planar models with different method and density of radiator filling (MN3–MN10). In addition, different conductive threads were used to create the samples, so that the tests carried out would be more comprehensive.

The first stage of the research consisted of measurements of the impedance of model antennas carried out in the frequency range from 0.5 to 1.2 MHz. The obtained results were analyzed within individual groups of samples: A (thread Syscom Liberator 40), B (thread Plug & Wear PW018A) and D (thread AMANN Silver-tech). The largest changes in impedance were observed between the MN1 and MN2 models, and the planar models, which is obvious, if only because of the differences in the amount of conductive thread used and the indirectly resulting changes in the values of parasitic parameters in the antenna. In the case of planar models, the observed changes in impedance between different embroidery patterns were not characterized by such a significant spread.

On the basis of the measured impedances of the model antennas, the conjugate match factor (CMF) was determined, which was then used to determine the bandwidth of the model antennas. CMF plots have much more similarity to each other (within a given group of samples and between groups) than impedance plots. A change in the resonant frequency relative to the impedance plots is observed—this is due to the impedance of the chip. Analysis of the bandwidth results obtained at this stage showed differences in the values of this parameter for different patterns, but these differences are in the order of 10%. A change at this level in many cases can be considered negligibly small.

The last stage of the research was to measure the read range of the RFID tag on the UHF band with model antennas. All the tested antennas met the basic requirement, i.e., the tag works in the entire band from 860 to 960 MHz (even though the CMF at the lower end of the band deviates significantly from 1), although in the specified frequency range, the value of the read range changes (the trend of changes coincides with CMF plots, which means that as the antenna matching changes, the read range changes as well). The results obtained could be improved if the impedance matching of the model antennas was improved.

An important observation regarding the obtained results is the change in the value of the frequency for which antenna matching is best (shift towards the upper limit of the analyzed band). This is due, among other things, to the parasitic capacitances found in the model antennas. This observation is consistent with the conclusions of other researchers, whose work is referred to in Section 1.1.

In Section 1.1, a hypothesis is presented that a change in the style and density of embroidery will cause a change in bandwidth, but a properly optimized embroidery process can reduce the amount of material used without significant deterioration of the parameters of the tested antenna. On the basis of the analyses carried out, this hypothesis can be considered confirmed. The stitch patterns used in the study had a negligible effect on changes to the bandwidth. Much more important in the context of the results obtained is ensuring the best possible impedance matching of the antenna.

5. Conclusions

Textile antennas are an important part of the growing new fields of the economy and are of interest to many researchers. In this paper, the authors analyze the effect of embroidery style on the bandwidth of the bow-tie antenna of a textronic UHF RFID tag. Different versions of planar antennas have been investigated; the degree of embroidery of the radiator depends on the style of embroidery. Analyzing the data collected during the experiments, it was found that there are noticeable differences in the value of the parameter of interest (antenna bandwidth), but these differences do not have a significant impact on the performance of the tag with which the model antenna is integrated.

Author Contributions: Conceptualization, P.J.-M. and M.N.; methodology, P.J.-M., M.N. and M.W.; validation, P.J.-M.; formal analysis, P.J.-M. and M.N.; investigation, P.J.-M. and M.W.; resources, M.N.; data curation, M.N.; writing—original draft preparation, M.N.; writing—review and editing, M.N.; visualization, M.N.; supervision, P.J.-M.; project administration, P.J.-M.; funding acquisition, P.J.-M. All authors have read and agreed to the published version of the manuscript.

Funding: This research received no external funding.

Institutional Review Board Statement: Not applicable.

Informed Consent Statement: Not applicable.

Data Availability Statement: All calculated and measured data will be provided upon request to the correspondent authors by email with appropriate justification.

Conflicts of Interest: The authors declare no conflicts of interest.

References

1. Younes, B. Textronics: A Review of Their Technological Aspects and Applications. *J. Text. Inst.* **2024**, *115*, 1509–1525. [CrossRef]
2. Younes, B. Smart E-Textiles: A Review of Their Aspects and Applications. *J. Ind. Text.* **2023**, *53*, 15280837231215493. [CrossRef]
3. Choudhry, N.A.; Arnold, L.; Rasheed, A.; Khan, I.A.; Wang, L. Textronics—A Review of Textile-Based Wearable Electronics. *Adv. Eng. Mater.* **2021**, *23*, 2100469. [CrossRef]
4. Meena, J.S.; Choi, S.B.; Jung, S.B.; Kim, J.W. Electronic Textiles: New Age of Wearable Technology for Healthcare and Fitness Solutions. *Mater. Today Bio* **2023**, *19*, 100565. [CrossRef] [PubMed]
5. Qu, Z.; Zhu, Z.; Liu, Y.; Yu, M.; Ye, T.T. Parasitic Capacitance Modeling and Measurements of Conductive Yarns for E-Textile Devices. *Nat. Commun.* **2023**, *14*, 2785. [CrossRef] [PubMed]
6. Lemey, S.; Rogier, H. SIW Textile Antennas as a Novel Technology for UWB RFID Tags. In Proceedings of the 2014 IEEE RFID Technology and Applications Conference (RFID-TA), Tampere, Finland, 8–9 September 2014.
7. Bouamra, W.; Osman, L. On-Body Investigation of Textile Antenna for Wearable RFID Applications. In Proceedings of the International Conference on Microelectronics (ICM), Sousse, Tunisia, 16–19 December 2018; Volume 2018.
8. Shah, A.A.S.A.; Abd Rahman, N.H.; Ali, M.T.; Ahmad, M.R.; Nordin, S.A.; Zaidi, N.I.; Abdul Aziz, N.H.; Yamada, Y. Design of Electro-Textile UHF RFID Tag through T-Match Network. In Proceedings of the 2018 IEEE International RF and Microwave Conference, Penang, Malaysia, 17–19 December 2018.
9. Jiang, Y.; Xu, L.; Pan, K.; Leng, T.; Li, Y.; Danoon, L.; Hu, Z. E-Textile Embroidered Wearable near-Field Communication RFID Antennas. *IET Microw. Antennas Propag.* **2019**, *13*, 99–104. [CrossRef]
10. Virkki, J.; Wei, Z.; Liu, A.; Ukkonen, L.; Björninen, T. Wearable Passive E-Textile UHF RFID Tag Based on a Slotted Patch Antenna with Sewn Ground and Microchip Interconnections. *Int. J. Antennas Propag.* **2017**, *2017*, 3476017. [CrossRef]
11. Lundquist, J.; Linkous, L.; Piper, M.E.; Sickey, Z.; Zimmet, K.; Mendoza, I.; Suresh, S.; Topsakal, E. Textile-Based Inkjet-Printed RFIDs: Exploring Wearable Antennas in the Real World [Bioelectromagnetics]. *IEEE Antennas Propag. Mag.* **2024**, *66*, 50–62. [CrossRef]
12. Dang, Q.H.; Chen, S.J.; Ranasinghe, D.C.; Fumeaux, C. Modular Integration of a Passive RFID Sensor with Wearable Textile Antennas for Patient Monitoring. *IEEE Trans. Compon. Packaging Manuf. Technol.* **2020**, *10*, 1979–1988. [CrossRef]
13. Luo, C.; Gil, I.; Fernandez-Garcia, R. Textile UHF-RFID Antenna Embroidered on Surgical Masks for Future Textile Sensing Applications. *IEEE Trans. Antennas Propag.* **2022**, *70*, 5246–5253. [CrossRef]
14. Luo, C.; Gil, I.; Fernandez-Garcia, R. Electro-Textile UHF-RFID Compression Sensor for Health-Caring Applications. *IEEE Sens. J.* **2022**, *22*, 12332–12338. [CrossRef]

15. Lee, H.; Tak, J.; Choi, J. Wearable Antenna Integrated into Military Berets for Indoor/Outdoor Positioning System. *IEEE Antennas Wirel. Propag. Lett.* **2017**, *16*, 1919–1922. [CrossRef]
16. Sathishkannan, P.; Singh, R.K.; Ray, K.P.; Singh, M. A Compact and Flexible Dipole Antenna to Be Integrated into Military Berets. In Proceedings of the 2022 IEEE Microwaves, Antennas, and Propagation Conference (MAPCON), Bangalore, India, 12–16 December 2022.
17. Chung, Y.; Berhe, T.H. Long-Range Uhf Rfid Tag for Automotive License Plate. *Sensors* **2021**, *21*, 2521. [CrossRef]
18. Wang, S.; Liu, Y.; Ye, T.T. "Unconventionally Shaped" Antenna Design for UHF RFID Tags. *Int. J. Antennas Propag.* **2021**, *2021*, 9965252. [CrossRef]
19. Kommey, B.; Addo, E.O.; Tamakloe, E. Systematic Design and Prototyping of a Low-Cost Passive UHF-RFID Transponder. *Eng. Rep.* **2023**, *5*, e12636. [CrossRef]
20. Hossain, S.S.; Karmakar, N. An Overview on RFID Frequency Regulations and Antennas. In Proceedings of the 4th International Conference on Electrical and Computer Engineering (ICECE), Dhaka, Bangladesh, 19–21 December 2006.
21. Kamili, J.B.; Bhattacharya, A. Design and Investigation of Cavity Backed Bowtie Antenna with Unidirectional Radiation Pattern Using Characteristic Mode Analysis. *Frequenz* **2024**, *78*, 323–334. [CrossRef]
22. Iliadis, L.A.; Boursianis, A.D.; Rekkas, V.P.; Zaharis, Z.D.; Koulouridis, S.; Goudos, S.K. Aperture-Coupled Bowtie Patch Antenna Design for Autonomous Driving Applications. In Proceedings of the 2023 International Workshop on Antenna Technology, Aalborg, Denmark, 15–17 May 2023.
23. Kumar, T.; Harish, A.R. A Modified Bow-Tie Antenna for Universal UHF RFID Application. In Proceedings of the ISAP 2018—2018 International Symposium on Antennas and Propagation, Busan, Republic of Korea, 23–26 October 2018.
24. Aoues, B.; Zaid, J.; Denidni, T.A. A Miniaturized Printed Bow-Tie Antenna for Passive RFID Applications. In Proceedings of the 2020 IEEE International Symposium on Antennas and Propagation and North American Radio Science Meeting, Montreal, QC, Canada, 5–10 July 2020.
25. Chopra, R.; Agarwal, A.; Lakhmani, R. A Broadband Planar Bow-Tie Antenna with Quasi-Isotropic Coverage. In Proceedings of the 2021 IEEE Indian Conference on Antennas and Propagation, Jaipur, India, 13–16 December 2021.
26. Kumari, S.; Awasthi, Y.K.; Bansal, D. Design a Bow-Tie Antenna for Multiband Applications. In Proceedings of the 17th INDIACom; 2023 10th International Conference on Computing for Sustainable Global Development, New Delhi, India, 15–17 March 2023.
27. Senapati, H.S.; Chongder, P.; Ajith, K.K.; Maiti, S. Design Of UWB Bow Tie Antenna For GPR Object Detection. In Proceedings of the 2021 Advanced Communication Technologies and Signal Processing (ACTS), Rourkela, India, 15–17 December 2021.
28. Gonçalves Licursi de Mello, R.; Lepage, A.C.; Begaud, X. The Bow-Tie Antenna: Performance Limitations and Improvements. *IET Microw. Antennas Propag.* **2022**, *16*, 283–294. [CrossRef]
29. Vital, D.; Zhong, J.; Bhardwaj, S.; Volakis, J.L. Loss-Characterization and Guidelines for Embroidery of Conductive Textiles. In Proceedings of the 2018 IEEE Antennas and Propagation Society International Symposium and USNC/URSI National Radio Science Meeting (APSURSI), Boston, MA, USA, 8–13 July 2018.
30. Maleszka, T.; Kabacik, P. Bandwidth Properties of Embroidered Loop Antenna for Wearable Applications. In Proceedings of the European Microwave Week 2010, EuMW2010: Connecting the World, Conference Proceedings—European Wireless Technology Conference (EuWiT), Paris, France, 27–28 September 2010.
31. Chen, X.; Liu, A.; Wei, Z.; Ukkonen, L.; Virkki, J. Experimental Study on Strain Reliability of Embroidered Passive UHF RFID Textile Tag Antennas and Interconnections. *J. Eng.* **2017**, *2017*, 8493405. [CrossRef]
32. Moradi, E.; Björninen, T.; Ukkonen, L.; Rahmat-Samii, Y. Effects of Sewing Pattern on the Performance of Embroidered Dipole-Type RFID Tag Antennas. *IEEE Antennas Wirel. Propag. Lett.* **2012**, *11*, 1482–1485. [CrossRef]
33. Vuohijoki, T.; Shaikh, A.; Merilampi, S.; Ihalainen, T.; Virkki, J. Punch-Needled Passive UHF RFID Tag Dipole Antennas—Design, Fabrication, and Initial Wireless Evaluation. In Proceedings of the 2023 8th International Conference on Smart and Sustainable Technologies (SpliTech), Split, Croatia, 20–23 June 2023.
34. Liu, Y.; Xu, L.; Li, Y.; Ye, T.T. Textile Based Embroidery-Friendly RFID Antenna Design Techniques. In Proceedings of the 2019 IEEE International Conference on RFID, Phoenix, AZ, USA, 2–4 April 2019.
35. Brechet, N.; Ginestet, G.; Torres, J.; Moradi, E.; Ukkonen, L.; Bjorninen, T.; Virkki, J. Cost- and Time-Effective Sewing Patterns for Embroidered Passive UHF RFID Tags. In Proceedings of the 2017 International Workshop on Antenna Technology: Small Antennas, Innovative Structures, and Applications, Athens, Greece, 1–3 March 2017.
36. Song, L.; Zhang, D.; Rahmat-Samii, Y. Towards Embroidered Textile Antenna Systematic Design and Accurate Modeling: Investigation of Stitch Density. In Proceedings of the 2019 United States National Committee of URSI National Radio Science Meeting (USNC-URSI NRSM), Boulder, CO, USA, 9–12 January 2019.
37. Simorangkir, R.B.V.B.; Yang, Y.; Esselle, K.P. Performance of Embroidered Higher-Order Mode Antennas with Different Stitching Patterns. In Proceedings of the 2017 11th European Conference on Antennas and Propagation (EUCAP), Paris, France, 19–24 March 2017.

38. Song, L.; Zhang, B.; Zhang, D.; Rahmat-Samii, Y. Embroidery Electro-Textile Patch Antenna Modeling and Optimization Strategies With Improved Accuracy and Efficiency. *IEEE Trans. Antennas Propag.* **2022**, *70*, 6388–6400. [CrossRef]
39. Seager, R.; Zhang, S.; Chauraya, A.; Whittow, W.; Vardaxoglou, Y.; Acti, T.; Dias, T. Effect of the Fabrication Parameters on the Performance of Embroidered Antennas. *IET Microw. Antennas Propag.* **2013**, *7*, 1174–1181. [CrossRef]
40. Pinapati, S.P.; Chen, S.J.; Brittain, J.; Caldow, A.; Fumeaux, C. Embroidered Ground Planes for Wearable Antennas. *IEEE Trans. Compon. Packaging Manuf. Technol.* **2022**, *12*, 1029–1039. [CrossRef]
41. Jankowski-Mihułowicz, P.; Węglarski, M.; Chamera, M.; Pyt, P. Textronic Uhf Rfid Transponder. *Sensors* **2021**, *21*, 1093. [CrossRef] [PubMed]
42. Ziobro, A.; Jankowski-Mihułowicz, P.; Węglarski, M.; Pyt, P. Investigation of Factors Affecting the Performance of Textronic UHF RFID Transponders. *Sensors* **2023**, *23*, 9703. [CrossRef] [PubMed]
43. Jankowski-Mihułowicz, P.; Węglarski, M.; Wilczkiewicz, B.; Chamera, M.; Laskowski, G. The Influence of Textile Substrates on the Performance of Textronic RFID Transponders. *Materials* **2022**, *15*, 7060. [CrossRef] [PubMed]
44. Jankowski-Mihułowicz, P.; Węglarski, M.; Pyt, P.; Skrobacz, K.; Karpiński, K. UHF Textronic RFID Transponder with Bead-Shaped Microelectronic Module. *Electronics* **2023**, *12*, 4873. [CrossRef]

sensors

MDPI

Article

Impact of Particular Stages of the Manufacturing Process on the Reliability of Flexible Printed Circuits

Andrzej Kiernich [1,2,*], Jerzy Kalenik [2], Wojciech Stęplewski [1], Marek Kościelski [1] and Aneta Chołaj [1]

[1] Łukasiewicz Research Network—Tele and Radio Research Institute, Ratuszowa 11, 03-450 Warsaw, Poland; wojciech.steplewski@itr.lukasiewicz.gov.pl (W.S.); marek.koscielski@itr.lukasiewicz.gov.pl (M.K.); aneta.cholaj@itr.lukasiewicz.gov.pl (A.C.)

[2] Faculty of Electronics and Information Technology, Warsaw University of Technology, Nowowiejska 15/19, 00-665 Warsaw, Poland; jerzy.kalenik@pw.edu.pl

* Correspondence: andrzej.kiernich@itr.lukasiewicz.gov.pl; Tel.: +48-22-590-7354

Abstract: The purpose of the experiment was to indicate which element of the production process of flexible printed circuit boards is optimal in terms of the reliability of final products. According to the Taguchi method, in the experiment, five factors with two levels each were chosen for the subsequent analysis. These included the number of conductive layers, the thickness of the laminate layer, the type of the laminate, the diameter of the plated holes, and the current density in the galvanic bath. The reliability of the PCBs in the produced variations was verified using the Interconnect Stress Test environmental test. The qualitatively best variant of the board construction was indicated using the signal-to-noise ratio and the analysis of variance method for each factor. The factors determined to be the most important in terms of reliability were the number of conductive layers and the current density in the galvanic bath. The optimal variant of the board construction was two conductive layers on a polyimide laminate, where the laminate layer was 100 μm thick, the hole diameter was equal to 0.4 mm, and current density was 2 A/dm^2 in the galvanic bath. Therefore, the plated experiment indicated the factors needed to obtain a high-quality product with a low failure rate.

Keywords: environmental tests; Interconnect Stress Test; design of experiment; Taguchi method; flexible printed circuit; product reliability; production process

1. Introduction

In the 20th century, Japanese engineer Genichi Taguchi said that "quality is the virtue of design" [1]. The principal challenge for engineers is to design and produce devices that meet the expectations and needs of the end users. Nevertheless, a task of equal importance is the production of a high-quality product. Any high-standard system not only functions properly as expected but also functions over a long period of time without any faults. It can be said that a robust product is reliable, useful, functional, easy to maintain, and efficient. With this in mind, it is possible to fully understand what Taguchi meant in the above quote. Furthermore, high-quality and reliable technology is important for further development of science and new innovations. Ambitious space programs and long-term scientific programs must be supported by the tools they use. Once again referring to Taguchi's philosophy, the product should not only be free of any defects when leaving the production line, but its high quality during its use should be ensured. The reliability of the device must be maintained regardless of operating conditions throughout its whole nominal period of use. In unfavorable circumstances, which can be called noise, the product should still meet the requirements. It is worth adding that designers are responsible for the robustness of the product. Regardless of how technologically advanced the factory is, mediocre designs will always result in mediocre products [1].

Unfortunately, flawed, low-quality products, with some even designed to fail over time, are becoming a bigger issue every day. This problem has two sides: ecological and economical. The issue of poor-quality products concerns all industries; however, herein, we focus on the electronics sector. According to the Sustainable Development Goals 2020 report issued by the United Nations, every year for the last decade, the average resident of our planet has produced from 5.3 to 7.3 kg of e-waste [2]. The amount of waste is constantly growing. Equally alarming is the fact that globally, less than 20% of e-waste is recycled. Raw materials such as copper, silver, gold, and rare-earth metals that are used in electronics are irretrievably lost. Moreover, every year, devices worth billions of dollars are thrown into the trash. This particularly affects the pockets of individual customers, who are forced to buy a new product because the previous one broke down very quickly.

To address the above issue, an experiment was conducted, which constitutes the basis of this paper. The aim of the research was to answer the following question: how do particular stages that occur in the production process affect the aging of electrical connections on flexible printed circuit boards (PCBs)? The connections were fabricated by metallizing holes. The Interconnect Stress Test (IST) was selected and used as an aging test during the experiment. This method made it possible to determine the resistance of the printed circuit board to rigorous assembly techniques or the operating conditions of an electronic device. The experiment was planned according to the Taguchi method, which allowed us to reduce the number of tests and determine the most favorable factors during the production process. The research results enabled the selection of optimized conditions and factors occurring in the process of manufacturing end products involving the Internet of Things and critical applications of Industry 4.0, including sensors.

Flexible printed circuit (FPC) technology has been present in the electronics industry for several decades. It is a response to the miniaturization and growing quality requirements of electronic equipment. Printed circuit boards made of flexible materials, such as polyimide and polyesters, offer design freedom in electronic devices, and save space, thereby reducing the size and weight of electronic equipment. Flexible laminates have another great advantage that distinguishes them from traditional, rigid glass–epoxy laminates. They solve the severe problem of the mismatch of coefficients of thermal expansion (CTE) of laminate and copper in plated through-holes (PTHs). Differences in the degrees of expansion of materials leads to excess stress generated in the components and electronic connections. In the worst case, this causes their breakage and circuit failure. However, the usage of flexible substrate, even if it is not fully plastic due to the assembly of components or the use of a connection with a rigid laminate, can reduce the effect of the mismatch of coefficients of thermal expansion while maintaining high reliability and low costs [3,4]. Flexible printed circuit boards have found application in relatively small devices such as notebooks, smartphones, and, thanks to its elasticity, wearable devices [5,6].

As was previously noted, the reliability of products should be a key requirement of their manufacture. At the prototyping and testing stage, this is checked in many ways. In the case of printed circuit boards, these are not just functionality tests. PCBs are often used in critical applications (in medical, space, military, or automotive applications and high-power devices) and work in demanding conditions (such as high-humidity environments or where mechanical shocks or rapid temperature changes occur). To check how PCBs behave in such conditions, they are subjected to many environmental tests. In order to ensure devices' safety, reliability, and compliance with the requirements and standards, products should successfully pass the analyses mentioned above. All tests should be performed in accordance with the IEC 60068 standard, which specifies a number of environmental testing methods. It specifies various atmospheric conditions for measurement to assess the ability of samples to perform under diverse scenarios [7]. These tests include accelerated HALT/HASS tests, water absorption tests (in a highly humid environment), vibration tests, and thermal shock tests. Each of them allows us to verify the device from a different perspective for particular assignments.

In this research, the Interconnect Stress Tests were selected as a methodology to check the reliability of flexible printed circuits. The IST, also known as thermal cycling induced by direct current testing, is an accelerated aging test method developed in the 1990s to measure changes in the resistance of plated through-holes and reliability of interconnection in printed circuit boards. They arose from the need for a fast, repeatable and reproducible inspection, the results of which would additionally correspond to those obtained from the existing methods. Moreover, the new technology was supposed to make it cheap and easy to analyze the results. The idea of the IST is based on the flow of direct current through the plated through-holes on the board. The current heats the metallization and the surrounding laminate. Eventually, as a result of the cyclic heating and cooling of the PCB, a failure occurs. The primary cause of the failure in ISTs is considered to be a mismatch in the coefficient of thermal expansion between the epoxy resin and the fiberglass in the substrate materials, and the copper in the through-holes. The effect of the failure is a crack in the metallization of plated holes. The IST relies on specially designed sample coupons placed on PCB fabrication panels [8–11].

Environmental tests, especially the IST, allow us to observe the behavior of individual technological solutions in high-power applications, such as GaN semiconductor technology [12]. The aim of these activities is, among others, the production of circuits with high electrical and thermal reliability for modern solutions that include GaN technology. This aligns with the global trends in the development of so-called Green Power Electronics. The efficient power supply and charging of electronic devices and electric vehicle batteries is now becoming a necessity. Combining the efficiency of the GaN with the resistance of flexible materials allows for significant increases in productivity and reliability.

This particular study investigated the reliability of flexible printed circuits by environmental thermal tests. This research included samples differing in their number of conductive layers, the type and thickness of the laminate used, the diameter of the metallized holes, and those made using different values of current density in the electroplating copper plating bath. The use of various production factors was aimed at determining which of them had the greatest impact on the reliability of the tested samples. The qualitatively best variant of the board construction was indicated using a signal-to-noise ratio (SNR) and an analysis of variance (ANOVA) method for each factor. To better understand the processes present in the tested PCBs, samples were analyzed by a metallographic cross-section, along with an observation under a scanning microscope and an infrared camera.

2. Materials and Methods

2.1. Selection of Experimental Factors

The production process of printed circuit boards is extremely complex. It consists of many stages and is influenced by many different factors. The path through which the PCB passes in the factory from the raw laminate to the finished product is shown in Figure 1. The first step in designing the experiment is to select several factors among all others. This is necessary due to the multitude of elements in the PCB production process and the limited time and material resources. In this paper, the focus was on factors that can meet the requirement of flexibility in the board and of elements of the PCB structure playing their critical roles during the Interconnect Stress Test. During the selection of the elements that would be a part of the experiment, decisions that were directly influenced by the designer of PCBs (or after a consultation with the producer) were taken into account.

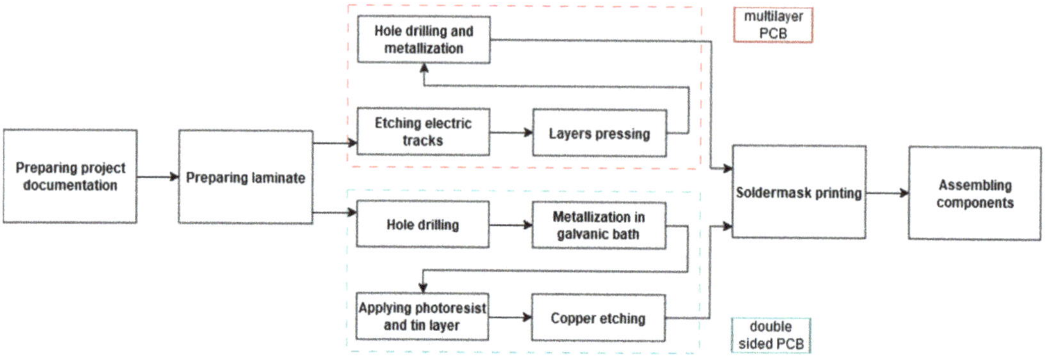

Figure 1. Scheme of production path of PCB.

Five factors that may have gad the biggest impact on the life of flexible printed circuits were chosen. These were as follows:

- Type of the laminate: polyimide and glass–epoxy (FR4) laminate;
- Number of conductive layers: 2 and 4 layers;
- Thickness of the laminate: 50 and 100 μm;
- Diameter of the plated holes: 0.3 and 0.4 mm;
- Current density in the galvanic bath: 1 and 2 A/dm^2—constant thickness of metallization equal to 20 μm.

2.1.1. Type of the Laminate

The first choice for the base material for FPCs is polyimide, a polymer plastic material, as it is one of the basic laminates used in the production of flexible printed circuits. This is due to the flexibility of the material and its durability when working at higher temperatures. However, a great disadvantage of this material is its high price [13]. In order to compare its performance in the thermal stress tests, we decided to analyze samples produced on glass–epoxy laminate in its most popular variant, FR4 [13]. The problem with glass–epoxy laminate is its rigidity. This is due to subsequent layers of glass fibers being stacked on one another. However, if enough thin base material is used, then the laminate should be flexible and meet the requirements of the application. Bearing this in mind, the purpose of comparing these two base materials was to investigate whether a traditionally used laminate would endure environmental conditions comparable to those of polyimide material.

2.1.2. Number of Conductive Layers

The number of conductive layers, which is two, among which copper plays the role of the conductor, is dictated by the nature of IST analysis. It is possible to design and fabricate plated through-holes only when the PCB has no less than two layers, where the conductor is located on both sides of the laminate. Currently, in advanced systems where integrated circuits with very numerous pinouts or a ball grid array (BGA) package are used, it is necessary to design multi-layer boards. Despite the increased complexity of the structure and the production process, this solution has many advantages [14]. The second simplest PCB design after the double-layer board is the four-layer board. Therefore, the reliability of the device was checked depending on the complexity of construction of a PCB.

2.1.3. Thickness of the Laminate

The most flexible films come in a narrow range of material thickness from 12 μm to 125 μm [3]. Hence, we decided to arbitrarily chose two values from the middle of the range, 50 μm and 100 μm. It is worth noticing that the practice shows that thinner laminates cause more problems during PCB production. It is necessary to achieve a reasonable tradeoff between the application requirements and the manufacturer's capabilities. The selection of the material's thickness as the next factor allowed us to check whether thin laminate is more resistant to thermal exposure. The thickness of a single laminate layer and the number of conductive layers together affect the total thickness of the PCB.

2.1.4. Diameter of the Plated Hole

The diameter of the plated holes varies a lot, from 100 μm to several millimeters. This is due to the many roles that PTHs play. They can function as an electronic interconnections between different parts of the circuit, heat sinks, and mechanical supports for component assembly, among many other purposes. As the size of the plated through-holes decreases, their durability also decreases. On the other hand, the reliability of the vias is highly influenced by the ratio of its diameter to its depth (the smaller the ratio, the more durable the connection). [15] It was also important to select commonly used hole sizes [16]. Bearing all this in mind, to maximize the potential of thermal stress testing, we decided to create holes measuring 0.3 mm and 0.4 mm.

2.1.5. Current Density in the Galvanic Bath

Shortening the production time of a single product is the top priority of every manufacturer. In the case of PCB production, one way to achieve this goal is by controlling the current density value during the electroplating process. The higher the current density, the faster the process of copper deposition [17]. Increasing the current density to the maximum possible value is a common practice in production plants. Thanks to this, the bath time can be made shorter, and a larger number of printed circuits can be produced. However, alternating the current density changes the crystal structure of the deposited metal. Because of this, the last element of the production process that was been taken into account when planning this experiment was the metallization of holes in the galvanic bath. This is due to the electrochemical deposition of a metal coating into a hole drilled into a laminate using a direct electric current. The coated surface serves as the cathode of the electrolytic cell. The anode is a block of a conductive material [13]. Copper is deposited by cation reduction. After consultation with the factory that produced the test samples, we decided to choose two metallization variants using a current density of 1 A/dm^2 and 2 A/dm^2. The aim was to find an answer to the question of whether the shape of copper crystals in the holes has an impact on reliability.

The selected factors and their levels are shown in Table 1 below.

Table 1. Factors taken into account during the experiment.

Factor	Levels	
	Level I	**Level II**
Type of the laminate	polyimide	glass–epoxy
Number of conductive layers	2	4
Thickness of the laminate	50 μm	100 μm
Diameter of plated hole	0.3 mm	0.4 mm
Current density in galvanic bath	1 A/dm^2	2 A/dm^2

2.2. Design of the Experiment

The experiment was designed and planned according to the method proposed by Genichi Taguchi. Designing experiments is a part of his entire approach to addressing the problem of product quality loss. The basic concept of Taguchi's philosophy is achieving a reduction in system instability at its output and therefore an improvement in the device's quality. When the observed characteristics at the output of the received product differ from the ideal or the desired ones, quality loss is considered to have occurred. With regards to quality, Taguchi introduces the classic signal-to-noise ratio (SNR) [18,19]. The signal power below is the desired response, and the noise power is the deviation from the ideal product characteristics:

$$SNR = 10\log_{10}\frac{\text{power of signal}}{\text{power of noise}}, \tag{1}$$

The best design is not one that maximizes the signal strength, but rather one that effectively reduces the impact of the noise. The SNR value should be as high as possible regardless of external factors that uncontrollably affect the system. Therefore, the goal of quality improvement efforts is to achieve a maximum signal-to-noise ratio. The aim of Taguchi's method is to reduce the variability in the final product under different conditions in the work environment. This ensures that the choices made in the laboratory also remain valid during production and use [18,19].

Considering the theory of quality loss, Taguchi proposes a way to plan the experiment. His idea is based on the partial–factorial design of experiments (DOEs) and the planning of subsequent studies based on an orthogonal array [18,19]. Thanks to this, Taguchi's method reduces the number of test runs required to perform. The orthogonal arrays are an extension of Latin squares. Written in the following form, $L_N(s^m)$, they are matrices of n rows by m columns. A specific property of orthogonal arrays is that in each pair of columns, each of the possible ordered pairs of elements appears the same number of times. The columns correspond to individual factors, the contents of cells in the columns correspond to factor levels and the rows correspond to test runs. Taguchi suggested a catalog of 20 arrays (including 18 orthogonal ones), but over time, the concept was developed and further suggestions were given. It is also possible to create submatrices by removing some columns from the array. In this way, many different multivariate experimental designs can be generated [20].

The experiment was conducted based on the selected criteria. For a full-factorial design of the experiment, $2^5 = 32$ tests would have been necessary to perform. We decided to design the experiment by using fractional-factorial design, and the Taguchi method was chosen for this purpose. This way, the number of tests required to perform the research was reduced. The first step in designing the experiment according to Taguchi is to construct an orthogonal array. As was explained in the previous paragraph, because 5 factors with 2 levels each were selected, we decided to use an $L_8(2^7)$ orthogonal array (array L_8 had 8 rows and 7 columns, so in case of the research described in this paper, 2 columns remained unused). Therefore, the entire experiment was carried out on the basis of 8 tests. The orthogonal array constructed for the purpose of the experiment is shown in Table 2. The 8 test runs, allowed for time savings, and this was achieved via the use of the fractional-factorial design for experiments according to the Taguchi's method, instead of the full-factorial design of experiments. This led to a huge reduction in costs in terms of time and material intake. In order to eliminate accidental errors that could have distorted the statistics obtained at the end of the study, from each produced PCB construction variant, three copies were drawn and tested within the experiment.

Table 2. Orthogonal array of designed experiment, where A—number of conductive layers; B—diameter of the metallized hole; C—type of laminate; D—laminate thickness; E—variant of the galvanic metallization process of holes.

Number of Experiment	A	B	C	D	E
1	2 layers	0.3 mm	polyimide	50 µm	1 A/dm^2
2	2 layers	0.3 mm	polyimide	100 µm	2 A/dm^2
3	2 layers	0.4 mm	FR4	50 µm	2 A/dm^2
4	2 layers	0.4 mm	FR4	100 µm	1 A/dm^2
5	4 layers	0.3 mm	FR4	50 µm	1 A/dm^2
6	4 layers	0.3 mm	FR4	100 µm	2 A/dm^2
7	4 layers	0.4 mm	polyimide	50 µm	2 A/dm^2
8	4 layers	0.4 mm	polyimide	100 µm	1 A/dm^2

2.3. *Design of Printed Circuit Board*

The tests performed for the purpose of this experiment required specifically designed and manufactured PCBs. Samples were not only fabricated in accordance with the selected factors but also had to be adequate for the IST. The IPC (formerly the Institute for Printed Circuits) standard number 650-TM 2.6.26 provides guidelines on how the samples should be designed. [21] The design of samples is shown in Figure 2, and an example of a PCB used as the sample is shown in Figure 3.

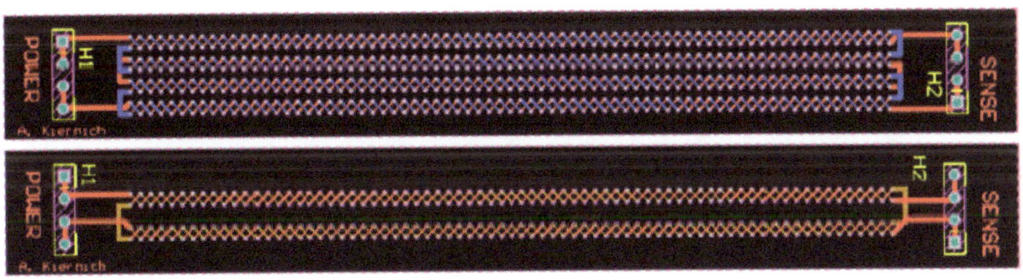

Figure 2. Design of samples, with the two-layer PCB above and the four-layer PCB below.

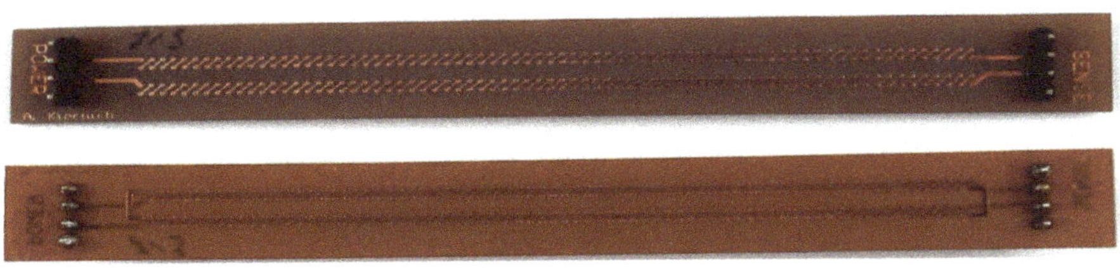

Figure 3. Example of a PCB used as the sample.

The standard distinguishes between two possibilities of a track routing. In the first method, named "Method A", there are two or more independent tracks on the test board. The power supply and heating circuit is marked as POWER, and the measuring circuit's resistance is marked as SENSE. In both methods, direct current flowing through the tracks heats the entire board to the set temperature, which then is cooled to room temperature. However, "Method A" was chosen because it allows for the measurement of the resistance in both the sensing and power supply circuits [21].

The basic design requirement for the samples was the dense arrangement of a large number of connections between the board layers. The aim of such a solution was to maximize the chance of a failure. The distance between successive metallized holes was 1.25 mm. The holes were planned along the entire length of the board. The power supply and measurement circuits were arranged to resemble a braid.

When designing the board, a decision was made to use a laminate with a 17.5 μm thick copper foil layer. In the technological process of producing the printed circuit mosaic, another copper layer measuring approximately 17.5 μm was applied to the tracks. This made the total thickness of a single track approximately 35 μm. The thickness of the metallization in the holes was about 20 μm as a result of electroplating. As was previously mentioned, one of the characteristic parameters of the electroplating process of copper is its duration. In the production of the test boards, previously selected current densities were used in two variants. At a current density of 1 A/dm^2 and with an efficiency of 100%, 0.2 μm of copper was deposited in one minute. For the established galvanic bath parameters, the deposition of the desired copper thickness took 110 min for 1 A/dm^2 and 55 min for 2 A/dm^2. As can be observed, increasing the current density significantly affects the production time.

Each PCB measured 14 mm × 120 mm. The standard recommends that the test sample has the shape of a rectangle, usually with dimensions of 0.6 and 5 inches, that is, respectively, 1.524 and 12.7 cm. They should have pinouts for the power supply and measurement circuits on both short edges, as well as male connectors with a spacing of 2.54 mm. [21] The thickness of the PCBs of different variants along with the rest of the information about test samples is presented in Table 3. In the case of the laminate made of polyimide, a material from the DuPont manufacturer (Wilmington, DE, USA) Pyralux AP series was used. Depending on the desired laminate thickness, AP8525R or AP8545R laminate was used (50 μm or 100 μm, respectively).

Table 3. Parameters of test samples.

PCB Parameter	Two-Layer PCB	Four-Layer PCB
Dimension	14 mm × 120 mm	14 mm × 120 mm
Thickness	120 μm or 170 μm (depending on variant)	305 μm or 405 μm (depending on variant)
Number of conductive layers	2	4
Number of plated holes	636	636
Total length of the traces	1299 mm	1331 mm
Interconnections	Top–Bottom	Top—1st signal layer 2nd signal layer—Bottom
Diameter of plated holes	0.3 mm or 0.4 mm (depending on variant)	0.3 mm or 0.4 mm (depending on variant)
Thickness of metallization in holes	20 μm	20 μm
Width of tracks	0.254 mm	0.254 mm
Thickness of tracks	35 μm	35 μm

Microvias

During the design of the test boards, the concept of used types of connections was clarified. In the case of a two-layer board, plated through-holes were used, while in four-layer boards, blind microvias were used. This design was chosen because it enabled us to test these two solutions and check them for durability and robustness.

A micro-hole, or microvia, is a blind one, and according to the definition of the IPC-T-50 standard, is a structure with a maximum aspect ratio of 1:1; it can have a depth of no more than 0.25 mm [22]. Blind micro-holes are characterized by a thinner layer of metallization near the neck of the microvia. This is the place where cracks most often occur. This happens because the expanding substrate presses down onto the neck of the microvia.

For this reason, producing PCBs with this solution gave us an idea of the difference between the robustness of microvias and of PTHs.

2.4. Performed Aging Tests

The entire IST procedure with details of individual stages and explanations are provided below.

The first step in the sample testing procedure was to measure the resistance of the POWER and SENSE circuits. Then, the exposure current was determined. The resistance values of the circuits on the boards differed slightly, which necessitated individual determinations of the exposure current for each sample. According to the IPC-TM-650 standard, 30-s tests, 60-s tests and pre-tests could be performed in order to adjust the exposure current. The 30-s and 60-s tests were used to pre-adjust the exposure currents to prevent the sample from heating up too quickly. Any number of 30-s and 60-s tests could be performed. After the initial estimation of the exposure currents, it was necessary to proceed to pre-tests. The sample was under full exposure for 3 min. If the sample reached the required temperature, the process of determining the exposure current was considered completed and the sample was accepted for subsequent cycles. The temperature reached in 180 s should be ±1 °C of the nominal value specified in the IPC standard [21].

The full exposure of samples could be progressively achieved after receiving a positive result from the pre-test. The maximum number of cycles was set to 10,000, and the possible relative change in resistance at the maximum and ambient temperature was set to 10%. The heating time in a single cycle was 180 s. After 3 min, the sample was cooled to room temperature by forced air. The cooling time was a function of the overall thickness and design of the board. Following the suggestion from the standard IPC-650-TM from 1999, the maximum cooling time was set to 2 min [11]. The resistance was measured every 2 s during heating in both circuits, POWER and SENSE; however, during cooling, it was only measured in the SENSE circuit.

When the change in the resistance crossed the threshold of 10% or the 10,000th cycle was reached, the process of sample exposure stopped. The measurement system recognized six different possible reasons for the test's termination. The first one was the successful completion of all planned exposure cycles. The remaining five determined reasons for interrupting the test were as follows: a break in the POWER circuit, a break in the SENSE circuit, the maximum temperature being exceeded, the dR (acceptable resistance change) at ambient temperature being exceeded, and dR being exceeded at the maximum temperature. Throughout the cycle, the system compared the current resistance with the value measured during the first cycle at the maximum and ambient temperatures. If the change in resistance during heating was more than ten times the initial value, the system considered it a break in the POWER or SENSE circuit. If the change was one and a half times greater than the acceptable value at the maximum temperature, it was treated as a temperature exceedance. However, if at the end of the heating or cooling process, the change in the resistance exceeded the acceptable value, the test was terminated due to the dR being exceeded at the maximum or room temperature.

The sample temperature was calculated based on the measured resistance using the following formula:

$$T = \frac{\frac{R_T}{R_{env}} - 1}{\alpha} + T_{env}, \tag{2}$$

where

- T—temperature of POWER or SENSE circuit;
- α—temperature coefficient (for copper it is equal to $3.9 \times 10^{-3}\ 1/K$);
- R_T—resistance at temperature T;

- R_{env}—resistance at room temperature;
- T_{env}—ambient temperature.

The summary of the IST procedure is presented in Table 4 below.

Table 4. IPC-TM-650 IST "Method A" review [21].

Test Parameter	Value
Temperature	150
Time of heating	3 min
Time at the maximum temperature	At least 1 s
Failure threshold	Change in resistance by 10%
Cooling method	By forced air
Resistance observation	Continuous
Temperature of sample	Calculation based on measured resistance

Workstation

The workstation was composed of the following elements:

- Control computer—laptop from DELL (Round Rock, TX, USA);
- Six-channel digital multimeter—3706A-NFP from Keithley (Cleveland, OH, USA);
- Power supply I—HMP2030 from Rohde and Schwarz (Munich, Germany);
- Power supply II—HMP2040 from Rohde and Schwarz (Munich, Germany);
- Temperature sensor—USB-Tset electronic thermometer from Aqua Lab (Warsaw, Poland);
- 12 V computer fan from Arctic (Brunswick, Germany).

The workstation is presented in Figure 4. The control unit was connected to other devices via a USB interface. The resistance in the SENSE circuit was measured with an ohmmeter in the range of 5 Ω with a resolution of 1 mΩ. During heating, the resistance of the POWER circuit was measured using the indirect method by reading the current and voltage (with a resolution of 1 mA and 1 mV, respectively), and then performing a resistance calculation. Eight samples were tested simultaneously. Each test board was cooled individually by its own fan, and each POWER circuit was individually powered from a programmable power supply. In order to minimize the impact of the external environment on the test result, the station was placed in an air-conditioned room, and the sample and the fan were covered with a screen.

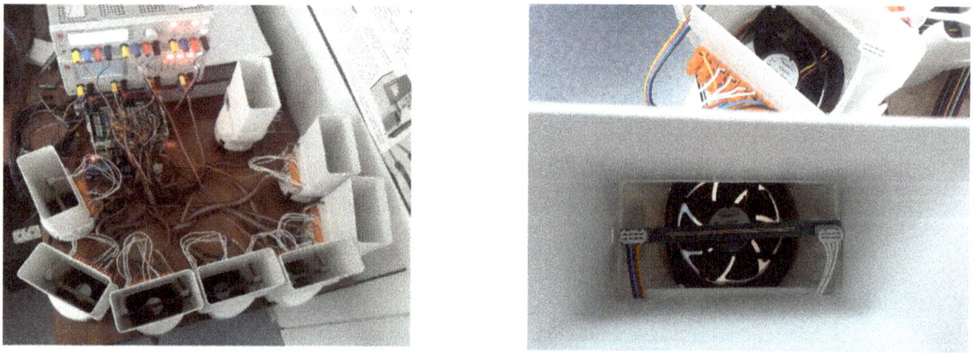

(a) (b)

Figure 4. Work station used during the experiment. (**a**) The whole work station; (**b**) a zoomed image of one testing unit.

3. Results

The Interconnect Stress Test results are presented in Figure 5. The test results are grouped into eight clusters that correspond to each of the board construction variants.

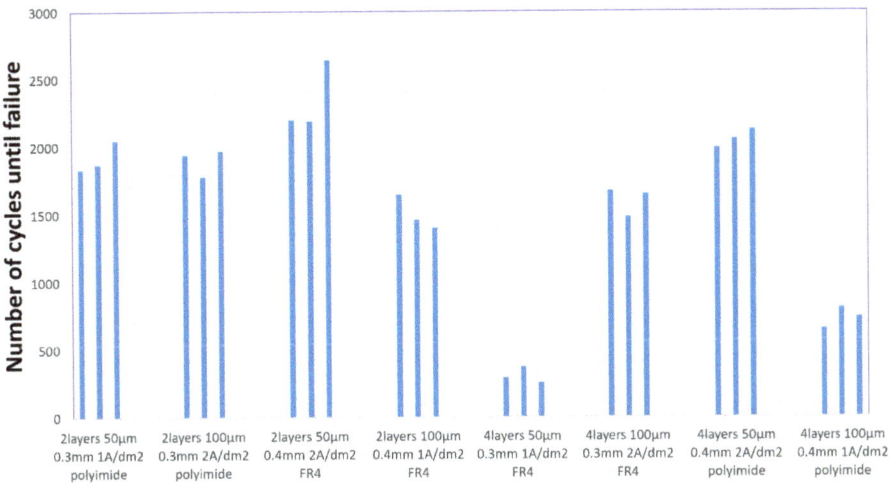

Figure 5. IST results.

Figure 5 shows the results of the experiment in an illustrative way. However, just by observing the graph itself, it can be seen that in the case of the variant of two-layer boards with plated through-holes, the number of cycles achieved was significantly higher than in the case of the construction with microvias. It can also be seen that the smallest deviation in results was achieved in the fifth experiment, while the largest was achieved in the third. Table 5 presents the basic statistical data. The arithmetic mean for each variant, independently and for the entire population, was calculated. The standard deviation for each variant was also provided, and the signal-to-noise ratio was calculated according to Taguchi's concept. In the conducted experiment, the more cycles the test board survived, the better it was; therefore, the formula for SNR is as follows:

$$\mathrm{SNR} = 10\log_{10}\left(\frac{1}{n}\sum_{i}^{n}\frac{1}{y_i^2}\right), \quad (3)$$

where

- n—number of samples;
- y_i—result of a sample number, i.

Table 5. Experiment statistics.

Number of Experiment	Arithmetic Mean	Standard Deviation	SNR [dB]
1	1917.33	93.43	65.62
2	1895.67	83.83	65.53
3	2340.67	210.27	67.29
4	1498.33	104.01	63.45
5	304.00	48.93	49.34
6	1598.00	86.73	64.03
7	2054.33	55.11	66.24
8	728.00	65.46	57.13
Mean	**1542.04**	**Sum**	**498.65**

An integral part of the analysis of the results is the calculation of the main effects, independent factors influencing the dependent variable. In the case of the conducted tests, this was the effect of the individual parameters of the production process on the durability of the test boards. Figures 6a, 7a, 8a, 9a and 10a contain tables of the main effect, their arithmetic mean values and the SNR coefficient for individual cases. The optimal parameter value is highlighted in bold. Figures 6b, 7b, 8b, 9b and 10b show illustrations of the analysis results above. They show the trajectory of the SNR parameter depending on the level of partial factors.

Number of layers	Arithmetic mean	SNR [dB]
2 layers	**1913.00**	**65.47**
4 layers	1171.08	59.19

(**a**)

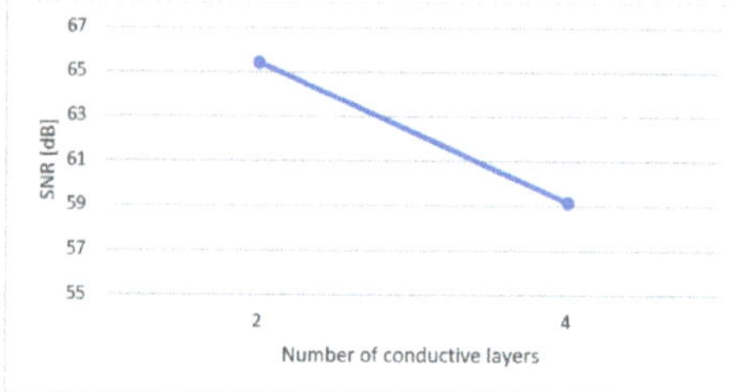

(**b**)

Figure 6. Main effect—number of conductive layers. (**a**) Table showing the following statistics: the arithmetical mean and SNR; (**b**) graph showing the trajectory of the SNR coefficient.

Thickness of laminate [μm]	Arithmetic mean	SNR [dB]
50	1654.08	62.13
100	**1430.00**	**62.54**

(**a**)

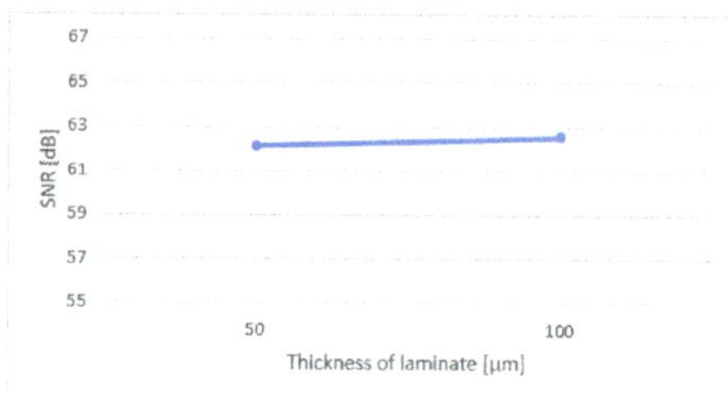

(**b**)

Figure 7. Main effect—thickness of laminate. (**a**) Table showing the following statistics: the arithmetical mean and SNR; (**b**) graph showing the trajectory of the SNR coefficient.

Type of laminate	Arithmetic mean	SNR [dB]
polyimide	**1648.83**	**63.63**
FR4	1435.25	61.03

(**a**)

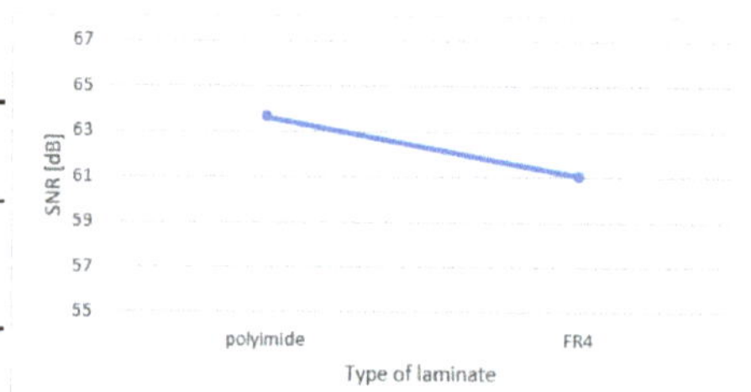

(**b**)

Figure 8. Main effect—type of laminate. (**a**) Table showing the following statistics: arithmetical mean and SNR; (**b**) graph showing the trajectory of the SNR coefficient.

Diameter of hole [mm]	Arithmetic mean	SNR [dB]
0.3	1428.75	61.13
0.4	**1655.33**	**63.53**

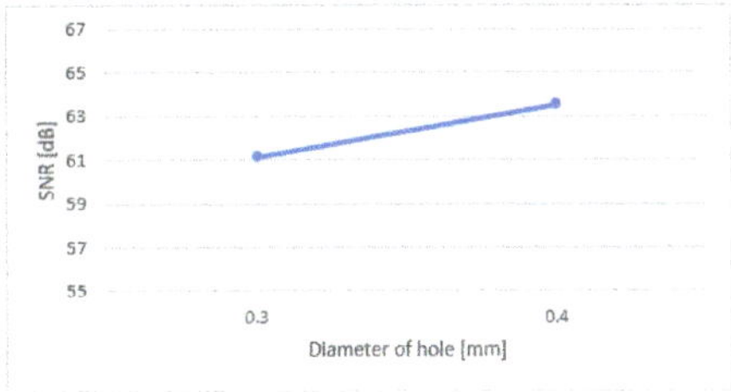

(**a**) (**b**)

Figure 9. Main effect—diameter of holes. (**a**) Table showing the following statistics: the arithmetical mean and SNR; (**b**) graph showing the trajectory of the SNR coefficient.

Current density [A/dm^2]	Arithmetic mean	SNR [dB]
1.0	1111.92	58.89
2.0	**1972.17**	**65.77**

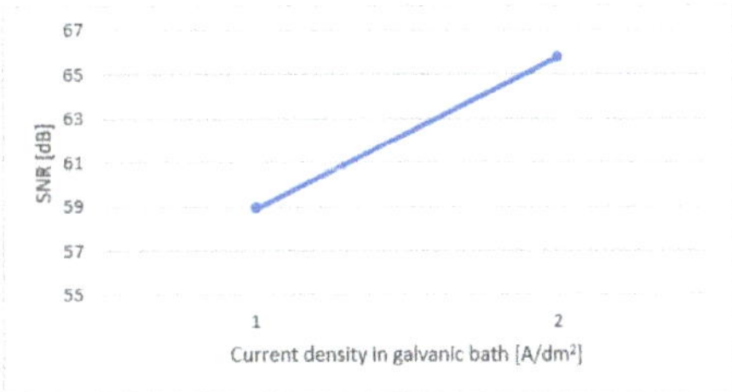

(**a**) (**b**)

Figure 10. Main effect—current density in the galvanic bath. (**a**) Table showing the following statistics: the arithmetical mean and SNR; (**b**) graph showing the trajectory of the SNR coefficient.

Based on the obtained results and the above data, it can be concluded that the optimal variant is a two-layer printed circuit board, manufactured on a polyimide laminate, with a laminate layer measuring 100 µm thick, a hole diameter of 0.4 mm and a current density of 2 A/dm^2 in the galvanic bath. The factors that have an upward trend are the laminate thickness, the diameter of holes and the current density in the galvanic bath. This means that the higher the value of a given parameter, the better the signal-to-noise ratio. In addition, the above graphs show that metallized through-holes and a polyimide laminate perform better than microvias and a glass–epoxy laminate.

In addition to the study proposed by Taguchi, a complex mathematical instrument, the analysis of variance, was performed. This tool allows researchers to assess the significance of differences between the means of individual groups. Conducting this type of analysis allows for a determination of the correlation between selected parameters of the production process and the reliability of flexible printed circuits. Moreover, ANOVA allows for a consideration of cases with many independent factors [23]. In the case of this research, an analysis was carried out for five factors. The calculation results are presented in Table 6.

For a significance level of 99%, the F statistic in all considered cases takes the value of 8.285 [24]. Based on the observation of the data from the table above, it can be concluded that the durability of the connections is equally influenced by two factors: the number of layers (type of connection) and the current density. For a lower significance level, only 90%, the F statistic has a value equal to 3.00698, as read from the table [24]. The final result is significantly influenced by two more factors: the laminate thickness and the hole diameter. The type of laminate does not have a significant effect on the final result.

Table 6. ANOVA table.

Source of Variation	Sum of Squares	Degrees of Freedom	Mean Square	F Value	P [%]
Number of layers	3,302,642.04	1	3,302,642.04	34.20	31.87
Thickness of laminate	301,280.04	1	301,280.04	3.12	2.91
Diameter of plated hole	308,040.04	1	308,040.04	3.19	2.97
Current density	4,440,180.38	1	4,440,180.38	45.99	42.84
Type of laminate	273,707.04	1	273,707.04	2.83	2.64
Error	1,738,007.42	18	96,555.97		16.77
Total	10,363,856.96	23			100

In the case of the most important factors, the number of conductive layers and the metallization warrant, the ANOVA confirms the results obtained after the analysis using the Taguchi method. Additionally, the percentage share of each factor in the overall scatter of observation values is given in column P. These values once again confirm the previous results.

3.1. Additional Research

The aim of additional tests carried out on test boards was to indicate the cause or the effect of the connection failures on the tested printed circuit boards. As part of the tests, photos were taken with a thermal camera, metallographic cross-sections were made and the surface of the metallized holes was observed using a scanning electron microscope.

3.1.1. Analysis by Thermal Camera

The temperature to which the sample heated up was checked using a thermal camera. This device images the recorded infrared radiation emitted by a tested object. This test served as a control. It also allowed us to observe how the dissipated heat was distributed on the test board. Below are thermographic images. Figures 11 and 12 show the sample during heating and the failure location. Thermal imaging camera A320 from Teledyne FLIR (Wilsonville, OR, USA) was used for the tests.

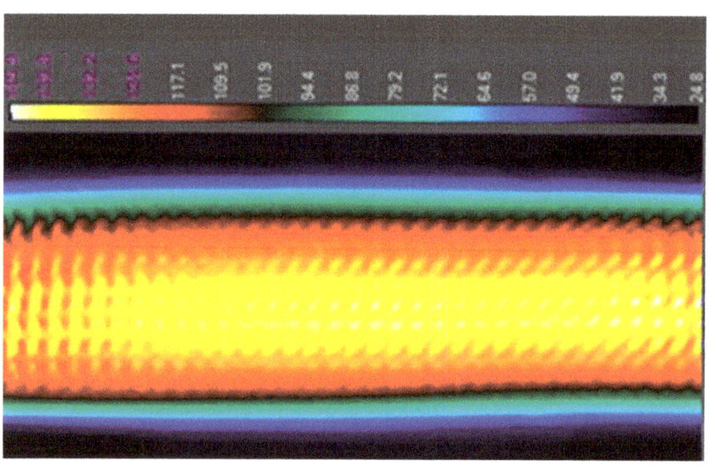

Figure 11. The test board during the heating cycle.

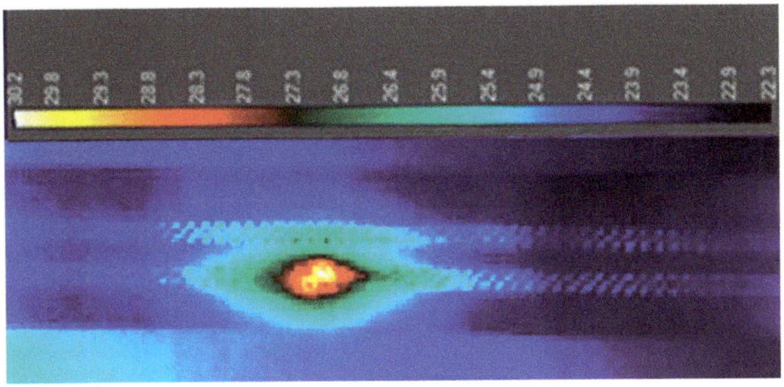

Figure 12. Location of the failure on the test board.

The first conclusion that comes to mind is that the highest temperature is reached in the middle of the board, as can be seen in Figure 11. The dissipated heat was distributed evenly along the entire length of the sample. The measurement performed with the thermal camera also made it possible to make sure that the measurement system was working properly and that the samples were heating up to the planned temperature. This examination also made it possible to locate the failure spot. The resistance increased in the place where the metallization was damaged. The thermographic image shown in Figure 12 clearly shows a hotter spot on the surface of the test board. Even at low exposure currents, with the current at a specific order of magnitude smaller current used during the tests, the difference was sufficient to select a place for a later cross-section to assess the degree of damage.

3.1.2. Analysis by Metallographic Cross-Section

In order to analyze the places of metallization cracks as a result of stresses caused by IST aging tests, metallographic cross-sections of selected test boards were performed. Based on the results of observations with a thermal camera, places were selected where there was a high probability of failure of the metallized connection occurring. Cross-sections were made for the indicated places. The samples for analysis were prepared on the MECAPOL P262 metallographic cross-section unit, Presi (Eybens, France), and the MetaServ 250, BUEHLER (Lake Bluff, IL, USA). Observations were carried out on the Eclipse L-150 metallographic microscope from Nikon (Tokyo, Japan), and the VHX6000 digital microscope from Keyence (Osaka, Japan). The analysis results are presented in Figures 13 and 14. The evaluation of metallographic sections was based on the criteria suggested by the IPC-A-600 standard [25].

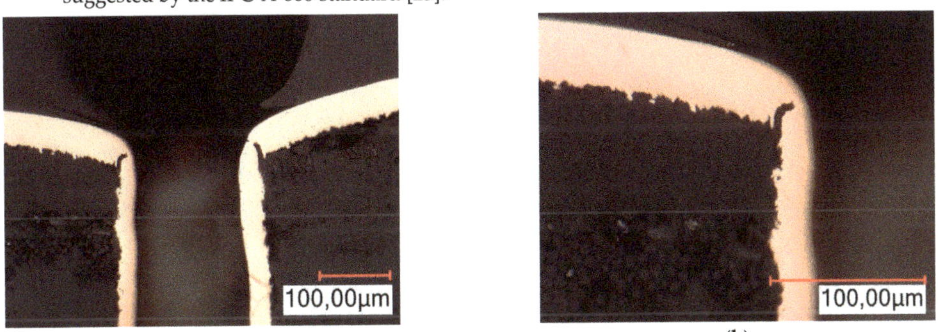

(a) (b)

Figure 13. Crack of metallization in PTH (**a**) metallographic cross-section—magnification ×700; (**b**) metallographic cross-section—magnification ×1500.

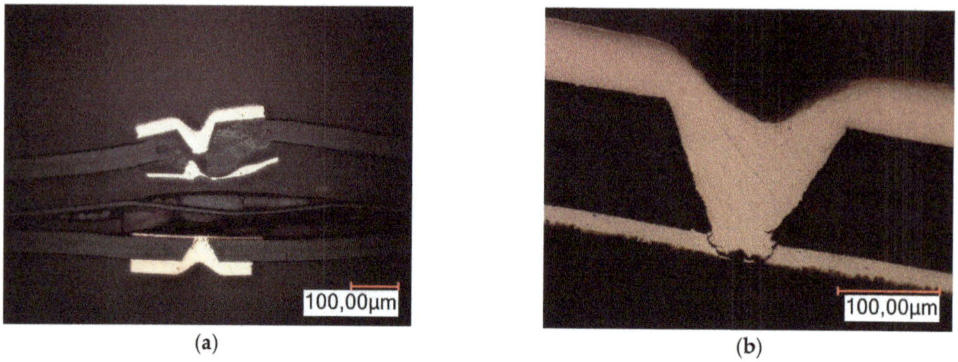

Figure 14. Crack of metallization in microvia (**a**) metallographic cross-section—magnification ×500; (**b**) metallographic cross-section—magnification ×1000.

All of the cross-sections clearly show the place of the crack and the damage to the metallization in the plated holes. Characteristic failure locations can also be observed on the cross-sections. For a PTH, this is the point of contact between the plated hole and the copper ring on the board surface. For a microvia, this is the so-called neck. The sections also show that in the case of PTHs, the crack starts from the "center" of the board. A similar observation could not be made for a microvia due to the complete damage to the connection.

In addition to the damage to the metallization resulting from the cracks, another effect of thermal stress and of failure can also be observed on the cross-sections. On the damaged boards, laminate delamination occurred, entailing the separation of individual components of the laminate from each other (fibers from resin or Kapton from glue). Voids and cracks in the laminate and the complete separation of individual components of the base material from each other can be observed in Figure 13a. In the case of PTHs, another defect occurred: lifted land above the board surface. A void appeared between the annular ring and the laminate. All the defects described above disqualify the printed circuits from further use.

3.1.3. SEM/EDX

Further tests were carried out using a scanning electron microscope, SEM JSM-7600F, with EDX spectrometry from Jeol (Tokyo, Japan). Cross-sections were made from two samples of PTHs with metallization obtained after bathing with a current density of 1 A/dm^2 and 2 A/dm^2, and then copper was etched using a solution of ammonium hydroxide with hydrogen peroxide. Cross-sections were made using a metallographic sample preparation system, type MECAPOL P262, PRESI (Eybens, France), and MetaServ 250, BUEHLER (Lake Bluff, IL, USA). Thanks to this, it was possible to observe the structure of individual copper layers on the laminate. Figure 15 shows layers of copper foil and electroplated with 1 A/dm^2. Figure 16 shows copper layers with the second current density variant. The tests were performed on samples that were not subjected to the thermal stresses.

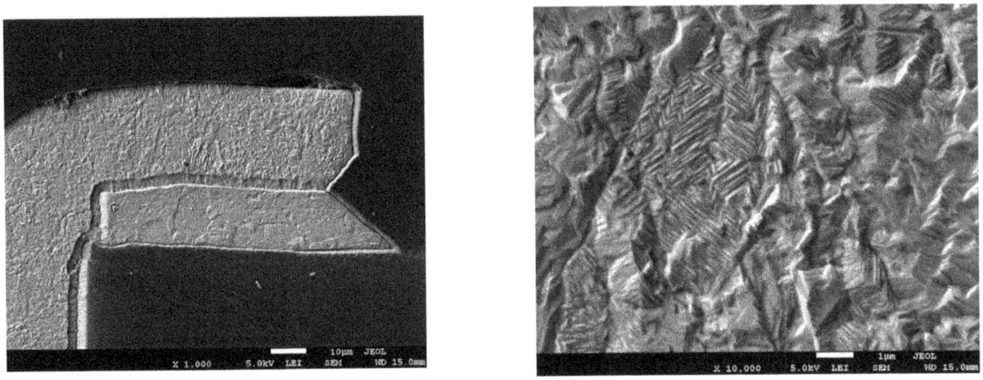

(a) (b)

Figure 15. Copper layers with metallization 1 A/dm^2: (**a**) magnification ×1000; (**b**) magnification ×10,000.

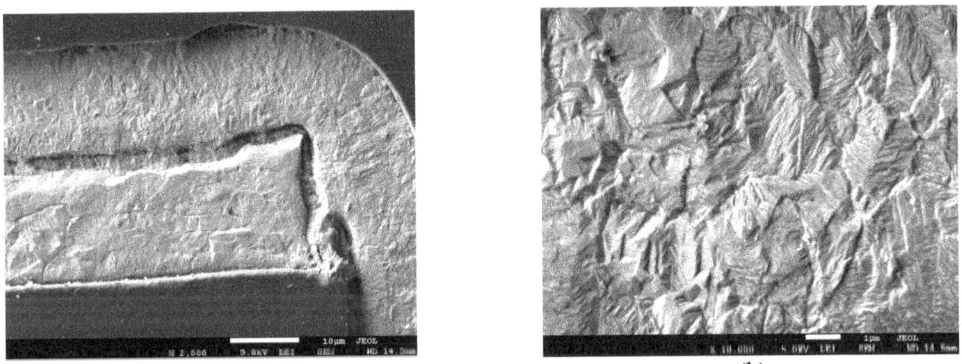

(a) (b)

Figure 16. Copper layers with metallization 2 A/dm^2: (**a**) magnification ×2000; (**b**) magnification ×10,000.

In the case of both samples, it is easy to distinguish between the copper that covered the plate in the galvanic bath and the copper foil that the manufacturer applied to the laminate. The latter has large crystals, with the same shape and structure on both samples. In the case of copper applied galvanically, the metal crystals are much smaller, creating a fine structure. The crystals of copper applied galvanically are several times smaller than those from copper foil. In the case of copper applied as a result of the galvanic bath, the difference in the size of the crystals for different variants is also clearly visible. The structure for the variant with a current density of 1 A/dm^2 is finer.

4. Discussion

After conducting the experiments and analyzing the results, the following summary was conceived for the factors considered in the experiment.

The factor that turned out to be the most statistically significant is the current density in the galvanic bath, meaning that this is the factor that had the greatest impact on the robustness of printed circuits. The samples with the variant 2 A/dm^2 had an SNR that was 6.88 dB higher than that for the second variant. The use of a higher current density in the galvanic bath turned out to be more advantageous for the quality of metallized connections. Although fine-grained crystals are more resistant to mechanical stress, they are characterized by higher resistance. This second parameter turned out to be crucial in terms of the IST.

The number of conductive layers also plays a significant role. After final decisions related to the construction of the test boards, a different type of connection was assigned to each variant of the parameter. The specificity of the PTH's and microvia's constructions determined the role that the number of layers plays in the reliability of printed circuits. The plated through-holes proved to be much more reliable than the microvias. The SNR coefficient for samples with PTHs was 6.29 dB higher than that for the second variant of connections. In future studies, an additional advantage would be to also conduct tests on buried holes.

The thickness of the laminate layer has a small effect on the quality of the product. Unexpectedly, it turned out that the thicker the laminate, the better. However, the difference was so small that it was almost imperceptible; the difference in SNR for two variants is equal to 0.41 dB. This may be due to the flexibility of the boards. For both thicknesses, 50 μm and 100 μm, the printed circuits were flexible, even though in the worst case, the entire board was 405 μm thick. Samples made of both polyimide and glass–epoxy laminates were not rigid. In the case of the ISTs, where the work of the material and its resistance to thermal expansion were important, the flexibility turned out to be a huge advantage. Despite the fact that, in the case of a thin laminate, the effect of the volume increasing with increasing temperature was smaller than that in the case of a thick material, the thicker laminate turned out to be slightly better. On the other hand, by referring once again to the previously discussed parameter, it should be noted that two-layer boards performed much better than four-layer ones. Their performance could have been influenced not only by the construction of the connection itself, but also by the thickness of the entire board.

The diameter of the holes also has a small effect on the final reliability. In accordance with the predictions, the larger holes had better results. However, it should be noted here that the obtained experimental results in many cases exceeded 1000 cycles. This is considered very high performance. In order to increase the failure rate of the test boards. the ratio of the hole diameter to its depth, that is, the aspect ratio, would have had to be reduced. Currently, this factor is from 0.125:1 to 0.3:1. In future studies, the aspect ratio should be reduced, for example, by using smaller-diameter holes. In the case of the samples analyzed during this research, the smallest metallized hole had a diameter of 0.3 mm. However, this value is close to the technology limit. Not many manufacturers are able to fabricate metallize holes with a diameter smaller than 0.1 mm. Not only is the drilling process problematic, but so is the metallization of the holes. For this reason, such small holes are difficult to implement. The advantage here is the small thickness of the laminate, which can facilitate the production process itself. Another way to reduce the shape factor is to increase the thickness of the base material. However, it should be noted that this can lead to the basic requirement for the board, its flexibility, not being met. Additionally, in the case of the microvias, their limitation is the maximum value of the aspect ratio, which was 1:1 in this study.

The type of the laminate used led to a similar difference in SNR coefficients to that in the diameter of the holes, about 3 dB, but according to the analysis of variance, this factor is insignificant. Polyimide, as a base material, performed better than the classic glass–epoxy. The small difference may have resulted from the previously discussed properties of the tested circuits, namely their flexibility.

The three factors mentioned above, the laminate thickness, the diameter of the metallized holes and the type of the laminate, did not play a major role in the conducted experiment. The tested printed circuits, with all structure variants, were relatively thin and retained their flexibility. These two properties of the test boards turned out to be of great importance for their reliability during the IST analysis.

In order to confirm the obtained results, an additional verification experiment was conducted. The test samples were manufactured based on the optimized factors. The printed circuit boards for the verification of the results were manufactured with two conductive layers placed on a polyimide laminate, where the laminate layer was 100 μm thick, the diameter of the holes was equal to 0.4 mm and a current density of 2 A/dm^2 was applied in the galvanic bath. The IST tests were performed for three samples. The obtained results are presented in Table 7 below.

Table 7. Results of the verification test.

	Sample 1	Sample 2	Sample 3	SNR
Verification test	2816	2586	2675	68.59

Based on the observation of the results obtained from the verification experiment and the previous ones, it can be concluded that the specified optimal production factors of flexible printed circuits have a positive and noticeable effect on the durability of the electronic connections. Ultimately, the goal of improving the quality of the product is to find the best values of the factors under control that occur in the production process in order to maximize the SRN coefficient. The signal-to-noise ratio for the verification experiment is the best among all those obtained in the conducted experiments. This research showed that the theory behind the design of experiments using the Taguchi method gives the desired effect, i.e., it indicates the optimal solution.

Two parameters turned out to have a major impact on the final result. More research is needed to verify whether the observed trends in the impact of the factors on the reliability will be maintained. For this reason, it is necessary to prepare tests and analyze the factor that has the greatest influence on the results, at three levels; in this study, this factor was the current density in the galvanic bath. In continued research on this topic, it is also worth investigating how holes with smaller aspect ratios will behave, using several variants. An experiment including three or more levels of a factor such as that above will provide a more complete picture of its effect on the robustness of flexible printed circuits.

5. Conclusions

Based on the evaluation of the results obtained from the experiment planned using the Taguchi method to examine the reliability of flexible printed circuits using the Interconnect Stress Test, the best variant of the board's construction was qualitatively determined, which ensured a high-quality final product with a low failure rate. The analysis was performed using the signal-to-noise ratio and an analysis of variance for each factor taken into account during the experiment. The factors that turned out to be the most important in terms of reliability are the number of conductive layers and the current density in the galvanic bath, rather than the type and thickness of the laminate, and the diameter of the metallized holes, which have a relatively small impact on the result. A higher value of current density in the plating bath will result in greater robustness. This research shows that the two-layer structure is more reliable than the four-layer structure, and that plated through-holes are more reliable than the microvias. The thickness of the laminate, the diameter of the vias and the type of base material did not have a significant impact on the test results, because the printed circuit boards in all considered cases were relatively thin and retained their flexibility. In the end, the optimal board construction variant was the use of two conductive layers, on the polyimide laminate, where the laminate layer was 100 μm thick, the hole diameter was equal to 0.4 mm and there was a current density of 2 A/dm^2 in the galvanic bath.

Based on the optimal factors, the samples for the verification test were produced. Based on the observations of the results obtained from the verification experiment and the previous ones, it can be concluded that the specified parameters of the production process of flexible printed circuits have a positive and noticeable impact on the durability of the connections. They therefore have an innovative impact on the quality of products, and high reliability opens up opportunities for their application in Internet of Things, the space sector, medical industries and Industry 4.0. Moreover, this research methodology shown the high durability of the particular PCB construction variants against thermal stress. The PTHs work well as heat sinks, which is important in the application of GaN technology in high-power semiconductor devices [12]. This experiment achieved the set goal and presented a number of factors that determine the quality of PCBs. The answer to the question of how to achieve high-quality and reliable electronic products while limiting the consumption of raw materials and reducing the amount of waste was obtained.

Author Contributions: Conceptualization, A.K., J.K., W.S., M.K. and A.C.; methodology, A.K., J.K., W.S., M.K. and A.C.; software, A.K.; validation, A.K. and J.K.; formal analysis, A.K.; investigation, A.K.; resources, A.K. and W.S.; data curation, A.K.; writing—original draft preparation, A.K.; writing—review and editing, A.K., J.K., W.S., M.K. and A.C.; visualization, A.K.; supervision, J.K.; project administration, A.K. and J.K.; funding acquisition, W.S. and M.K. All authors have read and agreed to the published version of the manuscript.

Funding: This research was co-financed by the President of the Łukasiewicz Center as a part of the project GANSUP—Opracowanie zasilaczy liniowych w technologii GaN do zastosowań w oświetleniu i sieciach typu smart grid (grant no. 1/Ł-IEL/CŁ/2022).

Institutional Review Board Statement: Not applicable.

Informed Consent Statement: Not applicable.

Data Availability Statement: The raw data of the experiments can be requested from the authors.

Acknowledgments: The presented research was part of the master's diploma thesis, titled "Wpływ czynników występujących w procesie produkcyjnym na niezawodność elastycznych obwodów drukowanych", presented in 2023 by Andrzej Kiernich and supervised by Jerzy Kalenik at the Warsaw University of Technology.

Conflicts of Interest: The authors declare no conflicts of interest.

References

1. Taguchi, G.; Clausing, D. Robust Quality, Harvard Business Review, January–February 1990. Available online: https://hbr.org/1990/01/robust-quality (accessed on 14 November 2024).
2. United Nation. *Sustainable Development Goals Report 2020*; United Nation: New York, NY, USA, 2020; ISBN 978-92-1-101425-9.
3. Fjelstad, J. *Flexible Circuit Technology*; BR Publishing: Seaside, CA, USA, 2006; ISBN 0-9667075-0-8.
4. Cheng, I.C.; Wagner, S. Overview of Flexible Electronics Technology. In *Flexible Electronics: Materials and Applications*; Wong, W.S., Salleo, A., Eds.; Springer: Boston, MA, USA, 2009. [CrossRef]
5. Benarrait, R.; Ullah-Khan, M.; Terrien, J.; Al Hajjar, H.; Lamarque, F.; Dietzel, A. A Flexible Double-Sided Curvature Sensor Array for Use in Soft Robotics. *Sensors* **2024**, *24*, 3475. [CrossRef]
6. Park, K.; Kim, M.P. Advancements in Flexible and Stretchable Electronics for Resistive Hydrogen Sensing: A Comprehensive Review. *Sensors* **2024**, *24*, 6637. [CrossRef] [PubMed]
7. *Standard IEC 60068-1:2013*; Environmental testing—Part 1: General and Guidance. International Electrotechnical Committee: Geneva, Switzerland, 2013.
8. Winco, K.; Yung, C.; Liem, H.M.; Choy, H.H.S.; Man, Y.W. Correlating Interconnect Stress Test and Accelerated Thermal Cycling for Accessing the Reliabilities of High Performance Printed Circuit Boards. *Packag. Manuf. Technol.* **2011**, *1*, 2005–2017. [CrossRef]

9. Stewart, D. Interconnect stress testing (IST)—An overview of its development and capabilities. *Circuit World* **2003**, *29*, 20–26. [CrossRef]
10. Cauwe, M.; Vandevelde, B.; Nawghane, C.; Van De Slyeke, M.; Coulon, A.; Heltzel, S. Challenges in introducing high-density interconnect technology in printed circuit boards for space applications. *CEAS Space J.* **2023**, *15*, 101–112. [CrossRef]
11. Standard IPC-TM-650:1999-Number 2.6.26. Test Methods Manual-DC Current Induced Thermal Cycling Test. IPC International: Bannockburn, IL, USA, 1999.
12. Bi, W.; Kuo, H.; Ku, P.; Shen, B. *Handbook of GaN Semiconductor Materials and Devices*; CRC Press: Boca Raton, FL, USA, 2017; ISBN 978-1-4987-4713-4. [CrossRef]
13. Raymond, C. *Handbook of Printed Circuit Manufacturing*; Van Nostrand Reinhold Company: New York, NY, USA, 1985; ISBN 978-94-011-7014-7. [CrossRef]
14. Shimoto, T.; Matsui, K.; Kikuchi, K.; Shimada, Y.; Utsumi, K. New high-density multilayer technology on PCB. *IEEE Trans. Adv. Packag.* **1999**, *22*, 116–122. [CrossRef]
15. Su, F.; Mao, R.; Xiong, J.; Zhou, K.; Zhang, Z.; Shao, J.; Xie, C. On thermo-mechanical reliability of plated-through-hole (PTH). *Microelectron. Reliab.* **2012**, *52*. [CrossRef]
16. Matisoff, B.S. *Handbook of Electronics Packaging Design and Engineering*; Van Nostrand Reinhold Company: New York, NY, USA, 1990; ISBN 978-94-011-7049-9. [CrossRef]
17. Schlesinger, M.E.; King, M.J.; Sole, K.C.; Davenport, W.G. *Extractive Metallurgy of Copper*; Elsevier: Amsterdam, The Netherlands, 2011; ISBN 978-0-08-096789-9. [CrossRef]
18. Freddi, A.; Salmon, M. *Design Principles and Methodologies From Conceptualization to First Prototyping with Examples and Case Studies*; Springer: Bolonia, Italy, 2019; ISBN 978-3-319-95341-0. [CrossRef]
19. Dehnad, K. *Quality Control, Robust Design and the Taguchi Method, Wadsworth & Brooks/Cole Advanced Books & Software*; Springer Book Archive: Pacific Grove, CA, USA, 1989; ISBN 978-1-4684-1474-5. [CrossRef]
20. HamzaÇebi, C.; Li, P.; Pereira, P.A.R.; Navas, H. *Taguchi Method as a Robust Design Tool. Quality Control-Intelligent Manufacturing, Robust Design and Charts*; IntechOpen: London, UK, 2020; ISBN 978-1-83962-498-8. [CrossRef]
21. *Standard IPC-TM-650:2014-Revision A-Number 2.6.26*; Test Methods Manual—DC Current Induced Thermal Cycling Test. IPC International: Bannockburn, IL, USA, 2014.
22. *Standard IPC-T-50-Revision N*; Terms and Definitions for Interconnecting and Packaging Electronic Circuits. IPC International: Bannockburn, IL, USA, 2021.
23. Rutherford, A. *ANOVA and ANCOVA: A GLM Approach*; John Wiley & Sons: Hoboken, NJ, USA, 2011; ISBN 978-0-470-38555-5. [CrossRef]
24. Statistics Online Computational Resource, F Distribution Tables. Available online: http://www.socr.ucla.edu/Applets.dir/F Table.html (accessed on 14 November 2024).
25. *Standard IPC-A-600:2010-Revision H*; Acceptability of Printed Boards. IPC International: Bannockburn, IL, USA, 2010.

sensors

Review

Novel Technologies towards the Implementation and Exploitation of "Green" Wireless Agriculture Sensors

Loukia Vassiliou [1,2], Adnan Nadeem [1], David Chatzichristodoulou [1,3], Photos Vryonides [1,4] and Symeon Nikolaou [1,4,*]

1 Electrical Engineering Department, Frederick University, Nicosia 1036, Cyprus; lvassiliou@ari.moa.gov.cy (L.V.); anadeem.msee19rimms@seecs.edu.pk (A.N.); eng.cd@frederick.ac.cy (D.C.); p.vryonides@frederick.ac.cy (P.V.)
2 Agricultural Research Institute, Aglantzia 1516, Cyprus
3 RF and Microwave Solutions Ltd., Dromolaxia 7020, Cyprus
4 Frederick Research Center, Nicosia 1036, Cyprus
* Correspondence: s.nikolaou@frederick.ac.cy; Tel.: +3-57-2239-4394

Abstract: This manuscript presents the use of three novel technologies for the implementation of wireless green battery-less sensors that can be used in agriculture. The three technologies, namely, additive manufacturing, energy harvesting, and wireless power transfer from airborne transmitters carried from UAVs, are considered for smart agriculture applications, and their combined use is demonstrated in a case study experiment. Additive manufacturing is exploited for the implementation of both RFID-based sensors and passive sensors based on humidity-sensitive materials. A number of energy-harvesting systems at UHF and ISM frequencies are presented, which are in the position to power platforms of wireless sensors, including humidity and temperature IC sensors used as agriculture sensors. Finally, in order to provide wireless energy to the soil-based sensors with energy harvesting features, wireless power transfer (WPT) from UAV carried transmitters is utilized. The use of these technologies can facilitate the extensive use and exploitation of battery-less wireless sensors, which are environmentally friendly and, thus, "green". Additionally, it can potentially drive precision agriculture in the next era through the implementation of a vast network of wireless green sensors which can collect and communicate data to airborne readers so as to support, the Artificial Intelligence and Machine Learning-based decision-making with data.

Keywords: precision agriculture; additive manufacturing; humidity sensors; RFID sensors; WPT; UAVs; energy harvesting; beamforming; battery-less wireless sensors

1. Introduction

The advances in technology and the need for sustainable agricultural practices are driving Precision Agriculture (PA) and Smart Farming (SF) [1] as depicted in Figure 1. The most widely used technologies involve the use of IoT devices, UAVs, AI, and big data analytics [2] to optimize crop production, water and nutrients allocation, and decision-making procedures. By integrating data from various types of sensors, farmers can precisely monitor and manage field conditions, leading to increased yields while ensuring minimized negative environmental impact. The potentials of PA and SF smart farming include improved efficiency, cost savings, and the ability to address global food security challenges through data-driven agricultural practices tailored to local conditions.

The deployment of agriculture wireless sensors [3] has revolutionized farming practices by providing real-time data which are crucial for optimal crop management. The most widely used sensors in agriculture are moisture/humidity [4,5] and temperature sensors [6,7]. Soil moisture sensors aid in precise irrigation scheduling, while temperature sensors help prevent frost damage and optimize growing conditions. Crop health sensors, often utilizing multispectral imaging, enable early detection of pests and diseases. Since many sensors are wireless and are widely scattered in large area fields, their precise location must also be monitored. For wireless IoT sensors [8,9], additive manufacturing [10] can provide an environmentally friendly biodegradable package [11]. In addition, additive manufacturing enables the rapid production of customized housing for wireless agriculture sensors tailored to specific environmental conditions. With 3D printing sensor enclosures, farmers can design shapes optimized for the sensing feature and the deployment conditions. Overall, it streamlines sensor deployment [12], maintenance, and innovation in modern farming practices.

Figure 1. Conceptual diagram of precision agriculture.

RF energy harvesting circuits [13] capture electromagnetic energy from ambient radio frequency signals, converting it into electrical power to sustain battery-less wireless IoT sensors [14,15]. Designated transmitters utilize the Wireless Power Transfer (WPT) [16] principle and emit radiation which is collected by wireless sensors with an Energy Harvesting feature. This additional feature allows the wireless sensor platforms to eliminate the need for physical connections or batteries. By harvesting RF energy from the surrounding wireless communication networks or designated transmitters, these circuits enable the continuous operation of sensors in various environments. This approach enhances scalability and reliability while reducing maintenance costs and environmental impact, making it ideal for IoT deployments and remote monitoring applications. It is possible for these WPT transmitters to be carried by UAVs [17] in order to secure a short distance between the transmitter and the ground-based sensors. This customized WPT transmitter can be added to already-used UAVs, which are already exploited in PA [18] through high-resolution imaging, that allow farmers to monitor crops with precision and detect issues such as nutrient deficiencies, pest infestations, or irrigation problems. UAVs can swiftly cover large agricultural areas, providing timely data and potentially being able to wirelessly charge a large number of IoT wireless sensors with Energy Harvesting capabilities.

The current manuscript presents methods of utilizing novel technologies such as (a) additive manufacturing, (b) RF energy harvesting, and (c) the use of UAVs as WPT transmitter carriers so as to support the implementation and exploitation of battery-less "green" wireless sensors. RF energy harvesting refers to a feature on the receiver's side to be able to collect RF energy using an antenna, convert it to DC, and use it at will. WPT refers to the use of a designated transmitter in order to wirelessly charge receivers with the Energy Harvesting feature.

This manuscript is organized into three technical sections, and Sections 2–4 describe specific uses of the aforementioned technologies. More specifically, Section 2 discusses the use of additively manufactured passive humidity sensors; Section 3 discusses the addition of the EH feature on dual frequency receivers using either the UHF or the ISM band; and Section 3 discusses an experimental setup for a UAV-carried transmitter that is used to wirelessly charge (WPT) ground-based sensors. Finally, Section 5 provides the conclusion on the possible uses of the discussed technologies in Precision Agriculture.

2. Additively Manufactured Passive Agriculture Sensors for Humidity Sensing

Soil humidity is one of the most important parameters that needs to be monitored and controlled for most agriculture crops. Therefore, the implementation of soil moisture sensors was set as one of the primary objectives of the ICARUS project. The use of additive manufacturing for the implementation of battery-less "green" sensors led to the design and manufacturing of passive humidity sensors where the sensing principle is based on the frequency shift that is caused by the wireless sensor's antenna as a result of the humidity variation. Two solutions were investigated. The first one was the use of a compact dipole antenna at 3.5 GHz, where the humidity-sensitive PEDOT: PSS material was used, and the second one was the use of a UHF RFID antenna embedded in soil, which exploited the effective dielectric constant change, associated with the surrounding soil's moisture.

2.1. A 3.5 GHz 3D-Printed Dipole

The implemented dipole [19] consists of two metallic spirals and the termination load consisting of an inkjet-printed PEDOT: PSS material which is sensitive to humidity changes. 3D printing was used to implement the carriers of the metallic spirals and the sensing material cage that was porous in order to allow for the penetration of the soil in the sensing material compartment. Figure 2 indicates the dipole's geometric characteristics. The PEDOT: PSS block implements a termination load for the dipole, and its effective impedance changes with different humidity levels. As a result, the effective resonance of the 3.5 GHz dipole shifts for different humidity levels, and therefore the humidity can be indirectly identified. Table 1 presents the association of the effective εr with the recorded and verified moisture percentage for two frequencies: the introduced one at 3.5 GHz, and the subsequent case where a UHF dipole at 915 GHz is used instead.

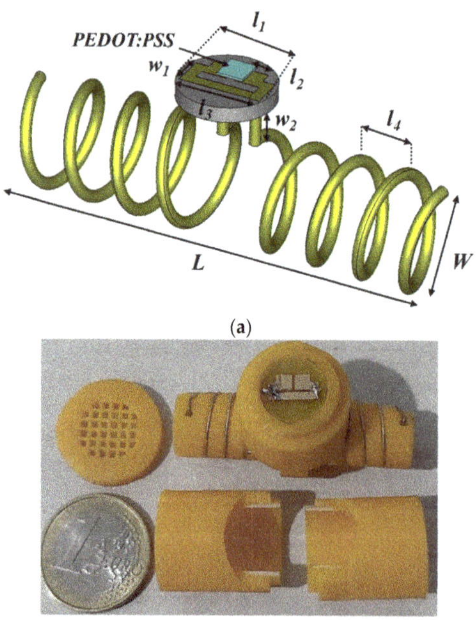

Figure 2. 3D-printed dipole for humidity sensing. Dimensions (mm): W = 15.1, L = 68, l_1 = 10, l_2 = 3, l_3 = 9, l_4 = 8.5, w_1 = 3, w_2 = 5, N = 3.5.

Table 1. The real part of the permittivity and loss tangent associated with the verified humidity percentage.

				For 3.5 GHz				
εr	**2**	**4**	**7**	**8**	**10**	**12**	**15**	**19**
tanδ	0.05	0.13	0.15	0.19	0.22	0.25	0.28	0.31
% of moisture	0	5	10	13	15	17	20	25
				For 915 MHz				
εr	**2**	**5**	**8**	**9**	**11**	**14**	**18**	**22**
tanδ	0.08	0.16	0.22	0.25	0.30	0.31	0.36	0.4
% of moisture	0	5	10	13	15	17	20	25

Figure 3 presents the frequency shifts of the 3D-printed dipole associated with the varying dielectric constant which is the result of different humidity levels. Despite the simplicity of the passive humidity sensing, there are two important drawbacks that limit its applicability and usefulness. The first one is based on the repeatability and reliability of the used method. Effective εr can be affected by other parameters as well, such as the salinity of minerals in the soil, and as a result, the detected frequency shifts cannot be exclusively associated with humidity changes. This can lead to inaccurate humidity readings. The second one is associated with the operation frequency. At 3.5 GHz, the propagation in soil is very lossy, and, consequently, the reading range of the wireless sensors is limited. In order to overcome the first drawback, the use of a more reliable IC-based digital sensor was preferred, and it is discussed in Section 3. To overcome the read range limitations, sensors operating at lower frequencies (under 1 GHz) were used.

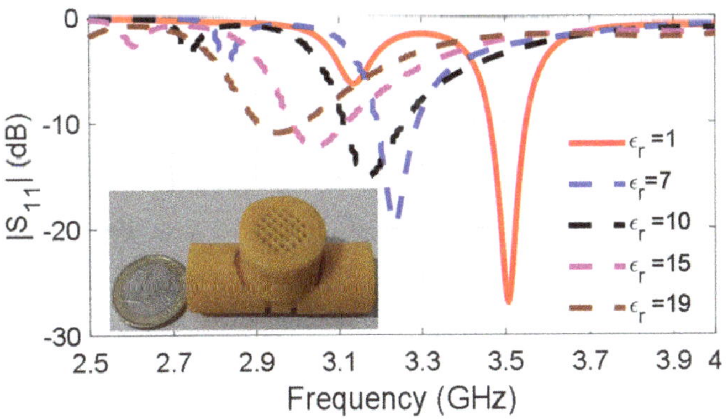

Figure 3. Self-resonance shifts in the presence of varying moisture soil.

2.2. UHF 3D-Printed Dipole

For the implementation of additively manufactured RFID-based sensors [19], a paper substrate and an inkjet-printed conductive nano-particle ink were used. The printed planar-folded dipole is depicted in Figure 4, and a 3D-printed protective cover is used to protect the sensitive paper substrate. The capacitive load of the IC has impedance 13.3-j122, and the folded arms of the dipole add inductive load for the conjugate matching. Self-resonance in free space occurs at 915 MHz and when the sensor is embedded in soil, the self-resonance shifts, as can be seen in Figure 5. The higher the effective dielectric constant, the lower the self-resonance occurs. The association of the effective εr with the moisture percentage has been presented in Table 1. Although the read range is improved as a result of the lower propagation loss at the UHF frequency, the reliability and repeatability issues for the readings remain, and this is why the use of IC humidity sensors has been preferred. To achieve high radiation efficiency, UHF radiators need to be relatively large (in the order of half-wavelength). Therefore, part of the research effort was focused on the miniaturization of the antennas of the wireless sensors.

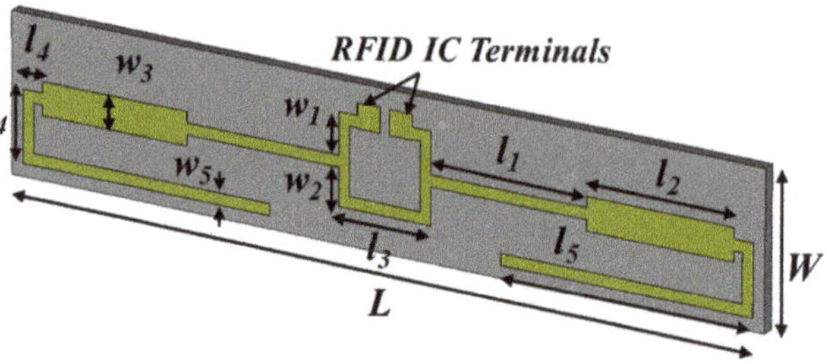

Figure 4. RFID dipole for humidity sensing. Dimensions (mm): W = 16, L = 96, l_1 = 20, l_2 = 19, l_3 = l_2, l_4 = 2.5, l_5 = 31, w_1 = 13.6, w_2 = 2.5, w_3 = 2.4, w_5 = 6.

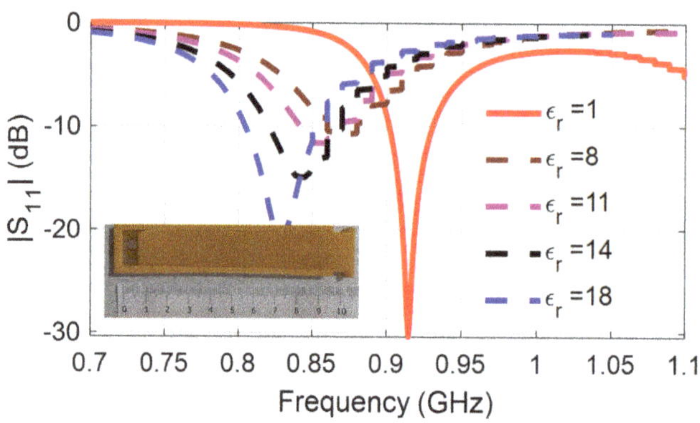

Figure 5. Self-resonance shifts in the presence of varying moisture soil.

3. Agriculture Sensors with Energy Harvesting Feature

For the extensive use of wireless sensors which are needed for Precision Agriculture, battery-less sensors should be used to avoid the cost and time consumption from battery replacement and possible soil contamination from the batteries' chemicals. PV cells are often used to power agriculture sensors, but this requires a large-sized sensor, and a large area and high-cost photovoltaic cell. It also requires direct exposure to the sun, and so, in other words, it cannot be used for sensors which are implanted in soil. An alternative possibility is the use of RF Energy Harvesting. Any radiation transmitter that is used for communication purposes can be also used as a potential source of RF power. An antenna, in combination with a rectifier (rectenna), can collect this RF power and convert it to DC power. This harvested energy can be used to power low-power sensors which act as transmitters for a short period of time, usually when interrogated by a designated reader, while they can remain in "listening mode" for a much longer period of time. During this period of time, they can harvest ambient RF power, or they can receive purposely transmitted wireless power in a designated Wireless Power Transfer (WPT) scheme so as to power the associated sensor. If the real-time converted power is insufficient for the needs of the sensors, a power management unit is also required. The concept of the harvested energy in a wireless power transfer scheme is presented schematically in Figure 6. Depending on the implementation, the wireless sensor may use a common antenna for the communication and for the Energy Harvesting system, or it may more often use dual antennas having the first designated for the communication system and the second for the Energy Harvesting system. In the current section, two implementations which use the UHF and ISM bands are presented.

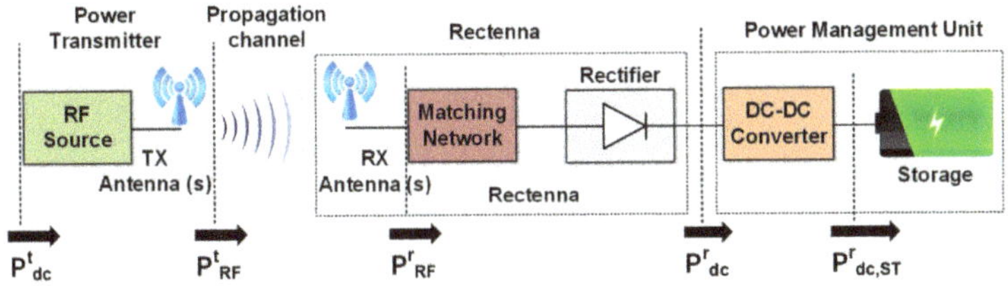

Figure 6. Wireless Power Transfer to a sensor with the Energy Harvesting feature.

3.1. ISM Sensor Platform with an Integrated UHF Energy Harvesting Feature

The dual-frequency system uses a meander UHF antenna [20] integrated with a lumped elements rectifier and a power management unit (PMU) based on the e-peas AEM30940 IC. For the communication module, a truncated-corners patch antenna with circular polarization is used which is connected to the TI CC2530 radio transceiver, integrating an Intel 8051 MCU, which adopts a protocol for communication named Simplicity. The TI-developed module is suitable for integration with a variety of digital sensors, including temperature and humidity sensors. The implemented wireless sensors platform with the integrated Energy Harvesting feature is presented in Figure 7.

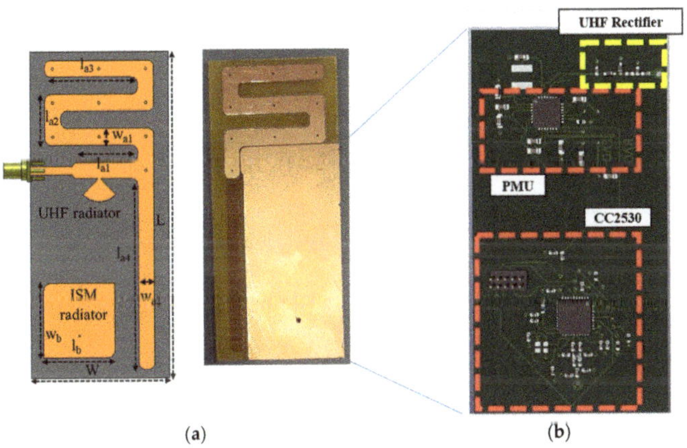

Figure 7. (**a**) Schematic of the modules' antennas. Dimensions (mm): L = 100, W = 47, la_1 = 18.4, la_2 = 16.5, la_3 = 37.5, la_4 = 62, lb = 24.5, lg = 75, wa_1 = 18.4, wa_2 = 18.4, wb = 24.5, wg = 35. (**b**) Prototype with the embedded circuit's implementation with the rectifier, the PMU, and the communication module in a single board.

A UHF meander monopole is used so as to reduce the overall size of the antenna. A radial stub is used to achieve the matching with the lumped elements rectifier. In order to avoid the large area required when distributed elements are used, lumped components are utilized instead. The equivalent circuit of the rectifier, with its return loss, and efficiency plots are presented in Figure 8. The effective RF-to-DC efficiency is defined as the ratio of the RF power at the input of the rectifier and the DC power on the ohmic load (R_L) that terminates the non-linear rectifier, as given in Equations (1)–(2). The RF power at the input of the rectifier is measured using a spectrum analyzer attached to the antenna that is connected with the rectifier, and the DC power on the load is calculated by measuring the DC voltage on the load using a digital multi-meter.

$$\eta = P_{RF}/P_L \quad (1)$$

$$P_L = V^2/R \quad (2)$$

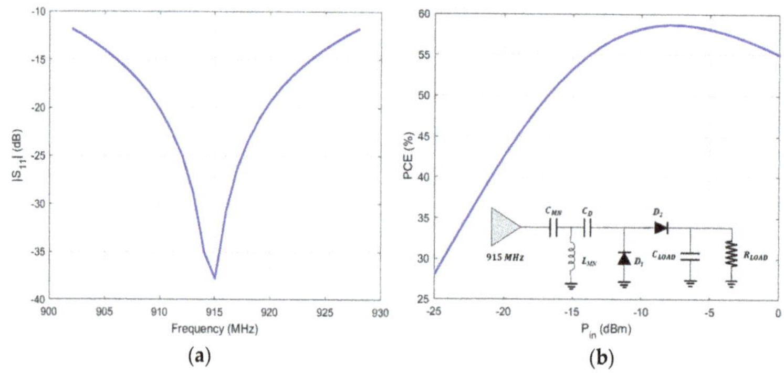

Figure 8. Lumped elements rectifier with insertion loss (| S_{11} |) and efficiency (PEC) plots.

The implemented rectifier is composed of a lumped input matching network (an RF capacitor CMN1 = 0.5 pF, and a shorted RF inductor in parallel LMN = 23 nH), two Schottky diodes D_1 and D_2 (Skyworks SMS7630-079LF), two RF capacitors at 1 nF (CD = CLOAD). The termination resistive load is 12.9 KΩ (RLOAD) and is setting the cut-off frequency of the output low-pass filter before the input stage of the subsequent PMU module.

For the implementation of the PMU e-peas AEM30940 IC has been selected. This IC can support a cold start from 380 mV which can be activated for a minimum input power equal to –22 dBm. After the cold start, assuming that the input voltage is between 50 mV and 5 V, the PMU-integrated boost converter can extract the available power from the source. This performance makes the specific lumped elements rectifier operational for even lower input power that can be as low as –32 dBm. This proposed PMU IC is able to supply low-voltages (up to 1.8 V) and typically drive a low-power microcontroller unit (MCU) like the used TI CC2530 IC, which is powered without the need of a battery source, providing a battery-less platform for agriculture sensors.

3.2. UHF Sensor Platform with an Integrated ISM Energy Harvesting Feature

An IC sensor for temperature and humidity-sensing capabilities is integrated into a similar platform where the UHF band is used for the communication module and the ISM band for the Energy Harvesting [21]. A PIFA antenna and a series rectifier operating at 2.45 GHz (ISM band) implement the Energy Harvesting unit that, in combination with the PMU, replaces the need for a battery. The PMU based on the e-peas AEM30940 IC provides a regulated 1.8 V DC voltage to the RFID tag IC (ROCKY100), assuming that the input RF power is higher than −22 dBm. For the sensing information communication, a UHF meander dipole antenna is used operating at 915 MHz. When the regulated 1.8 V is supplied from the PMU, the tag IC operates in semi-passive mode, and it effectively increases its read range. The used RFID IC is ROCKY100, which is EPC C1G2 compliant and is compatible with the power harvesting modules and SPI communication. This makes it compatible with the PMU and the HTS221 humidity and temperature IC sensor, which can be used for agriculture applications that involve soil measurements. The conceptual module and the IC layout of the proposed UHF platform with the integrated Energy Harvesting feature can be seen in Figure 9. The process of measuring the relative humidity and temperature is controlled with a Texas instrument mixed-signal microcontroller including two SPI interfaces to communicate with the RFID IC. Upon receiving an SPI-directed read request from the RFID reader, the ROCKY100 SPI bridge requests the value of the last measurement. Then, the recorded reading from the HTS221 IC is sent to the RFID IC, and from there to the RFID reader. The use of the integrated Energy Harvesting feature results in the enhancement of the read range of the wireless sensor and a battery-less, and thus “Green”, operation.

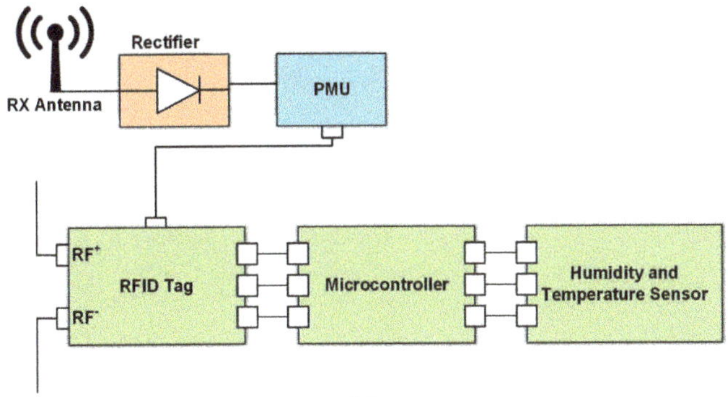

(a)

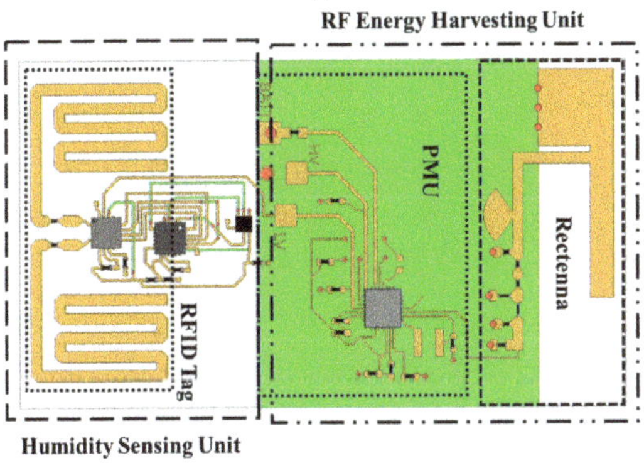

(b)

Figure 9. Block diagram and IC layout of the implemented module.

The commercially available HTS221 is a sensor for relative humidity and temperature sensing that includes a sensing element and can operate over a temperature range of −40 °C to +120 °C. The sensor IC requires a low supply voltage that can be covered from the low-voltage output terminal of the PMU, which is regulated at 1.8 V. The sensing element is integrated with an IC able to communicate the sensed information in the form of a digital signal to the microcontroller through the I²C/SPI (Inter-Integrated Circuit)/(Serial port interface) interface. Considering that a commercially available temperature and humidity IC sensor is used, emphasis is given to the implementation of the Energy Harvesting module and the UHF communication antenna. The two antennas, are presented in Figure 10 with their measured insertion loss.

A voltage-doubler topology is used as a rectifier. Despite its inherent non-linear nature, it is well-matched for different levels of input power ranging from −20 dBm to 0 dBm, as can be seen in the response in Figure 11. The effective RF-to-DC efficiency depends heavily on the input power, and it has a peak value at almost 60% for input power at 0 dBm (Figure 11).

The presented battery-less RFID-based humidity and temperature sensor with the integrated Energy Harvesting feature presents a "green" solution for agriculture applications. The sensing module utilizes the harvested power from the PMU to power up the ROCKY100 RFID IC. With the provided power, the IC operates in semi-passive mode, and it significantly increases its read range. The sensing information is communicated as a digital signal from the digital capacitive HTS221 humidity sensor using the SPI and I2C interface pins, and the data are transferred to the RFID IC through the MSP430 ultra-low-power microcontroller. When an interrogation read command is received, the sensor data are transmitted to the RFID reader, along with the RFID tag ID, through the RFID UHF meander dipole.

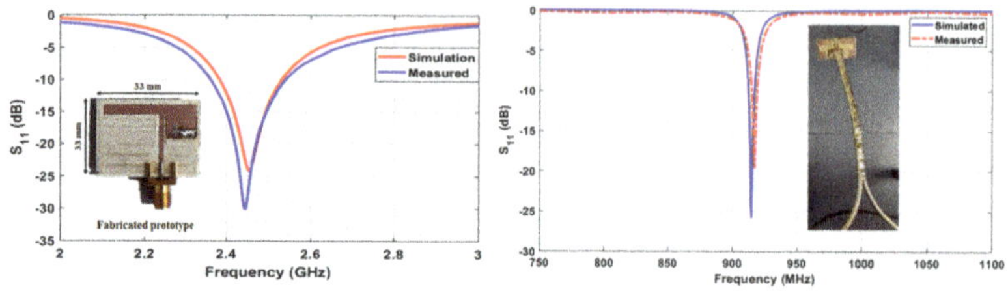

(a) (b)

Figure 10. The PIFA antenna for the Energy Harvesting ISM module, and the meander monopole for the communication UHF module.

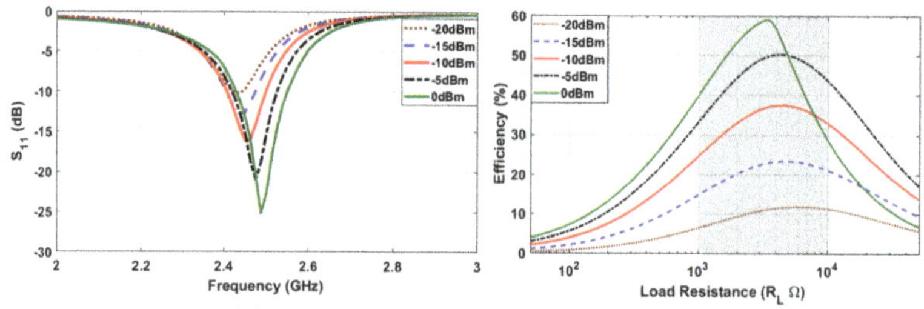

(a) (b)

Figure 11. Matching and efficiency of the implemented voltage doubler rectifier.

4. Wireless Power Transfer from UAV-Carried Transmitters to Ground-Based Sensors

The Energy Harvesting feature of the green wireless sensor platform is utilized when a transmitter is designated with the wireless charging of the sensors. This can be accomplished with Wireless Power Transfer, which is the radiation of un-modulated waves (power) directed in high-gain beams aiming to increase the power density at the receivers. Considering the free-space losses, it is often critical to maintain a small distance between the transmitter and the receiver modules. Carrying the transmitter on a flying UAV provides some controlof the distance between a mobile transmitter and a stationary ground-based receiver.

4.1. WPT at UHF Frequency–915 MHz

Wireless charging from a UAV-carried transmitter towards ground-based sensor platforms with the Energy Harvesting feature has been tested. In the described experiment, the block diagrams of both the transmitter and receiver modules are presented in Figure 12. The ground-based sensors use the UHF Energy Harvesting system that was described in Section 3.1 with a resonance frequency of 915 MHz (Figure 8). During the experiment, which took place in an open field, the UAV s hovered at a height of 1.6 m above the ground and the ground-based sensors were positioned in one of the 25 positions of the grid (5×5). During the WPT process, a direct line of sight (LOS) was ensured, and the voltage of the capacitor was monitored until it reached the maximum possible voltage at 3.5 V, which is equivalent to the maximum stored energy for the Energy Harvesting system. The measurement across the capacitor was repeated 25 times, one for every possible position of the sensor on the grid. Three different ground-based sensors were used, and the measurements implied that their charging was uncorrelated, indicating that the charging of multiple sensors can take place simultaneously.

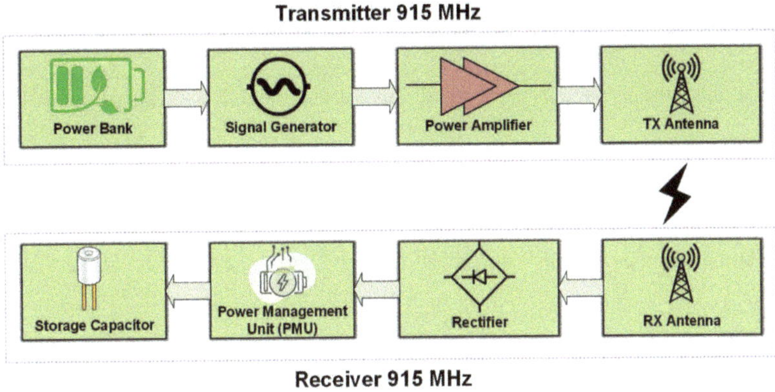

Figure 12. WPT experiment—TX and RX building blocks.

For the experimental setup of wireless power transfer from a UAV-carried Transmitter (TX) to the ground-based sensors, a frequency of 915 MHz (UHF) was used over the ISM so as to limit the free-space losses. In order to integrate the entire TX system on a flying UAV, a portable signal generator was used that produces 0 dBm output power at 915 MHz and was powered by a light-weight power bank. For the purpose of boosting the TX power, a power amplifier at 915 MHz was utilized that had 30 dB gain. For the transmitter antenna, a compact patch with gain 2 dBi was used.

On the receiver (RX) side, the compact circularly polarized antenna with gain 2 dBi was used in combination with a voltage doubler rectifier at 915 MHz. The rectified voltage was stored in a commercially available Power Management Unit (PMU) with a 4.7 mF capacitor as the power storage element.

For the experimental setup, a UAV with the transmitter hovered at a height of 1.6 m from the ground. One Energy Harvesting receiver (Figure 11) was placed on the surface of the ground in each of the 25 grid points (Figure 13), and the voltage on the 4.7 mF storage capacitor was monitored until it reached the maximum output voltage from the PMU. For the measurements, a direct line of sight (LOS) between the transmitter and the receivers was ensured.

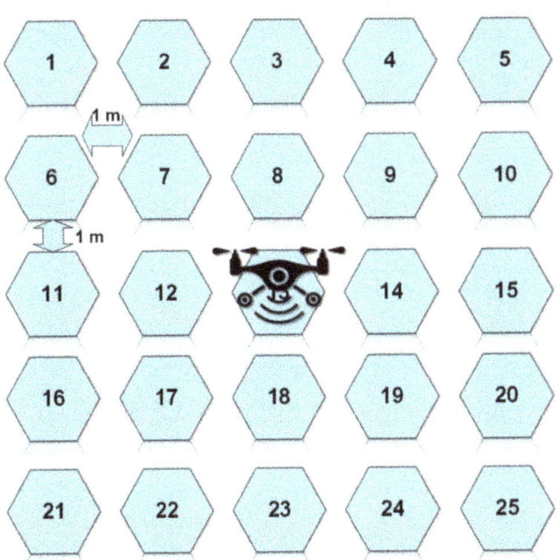

Figure 13. A 5 × 5 square grid for the experimental setup.

There are six cases presented in Table 2, where the Energy Harvesting receiver was placed at six different positions which are numbered on the rectangular grid on the ground. For each case and considering that the UAV was hovering at a fixed position of 1.6 m over square 13, the elevation and azimuthal angles are marked. The charging times which are recorded indicate the time that is needed, under the aforementioned conditions, for the 4.7 mF storage capacitor to reach 1.2 V, 1.8 V, 3.3 V, and 3.5 V, respectively. These voltage levels are associated with the output voltage, which can be provided by the PMU. According to the PMU specifications, the maximum voltage that can be supplied to the storage element is 4 V. The storage capacitor provides the necessary power to the sensing board that is connected to the PMU when there is no RF power available. The graphs of all six cases of energy storage (mJ) in a charging capacitor versus charging time (mins) are plotted in Figure 14. Expectedly, the closer a rectenna is the higher level of power it receives and, consequently, the faster the charging element acquires sufficient energy to power the communication module. In conclusion, the experimental setup indicated that the implemented Energy Harvesting module can provide sufficient power to allow for the battery-less operation of the used sensors, and for the given transmitter specifications, the RF energy harvesting was possible for a maximum distance of 18 m under optimum antenna orientation and polarization conditions. The duration of the WPT process depends on the flight duration and the flight altitude of the UAV that carries the UHF transmitter.

Table 2. Charging time for ground-based sensors with the EH system, which are physically located at various distances and in different angles compared to the UAV-carried WPT transmitter.

Grid No.	Distance (m)	θ_{angle} (degree)	φ_{angle} (deg)	Charging Time (mins)			
				(1.2 V) 3.38 mJ	(1.8 V) 7.61 mJ	(3.3 V) 25.6 mJ	(3.5 V) 28.8 mJ
13	1.6	0	0	3.5	5.5	8	9
14	1.87	0	30	4	6.5	10	12
15	2.56	0	50	5.5	8.5	17	19
10	2.75	25	55	5.5	8	15	19
9	2.14	45	40	5	7	13	14
5	3.25	45	60	7	10	18	22

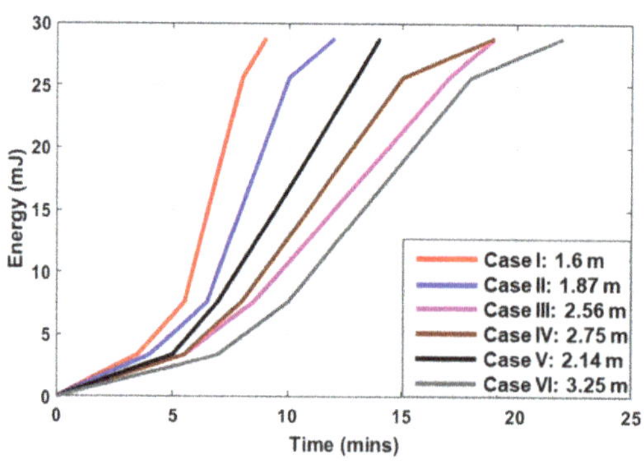

Figure 14. Charging time for a 4.7 mF capacitor to reach the maximum stored energy of 28.8 mJ (under the maximum voltage of 3.5 V) for different distances between the transmitter and the Energy Harvesting system.

4.2. *Communication Range between the UAV-Carried RFID Reader and the UHF Communication Module*

The block diagram of the UAV-carried communication system is presented in Figure 15. The communication module consists of three sub-systems. The WPT system has been described in the previous section. There is also a portable UAV reader for the communication with the RFID-enabled sensors, as well as a module that communicates with a base station over WiFi. This is implemented using an Arduino UNO microcontroller board with an ESP8266 Wi-Fi communication module.

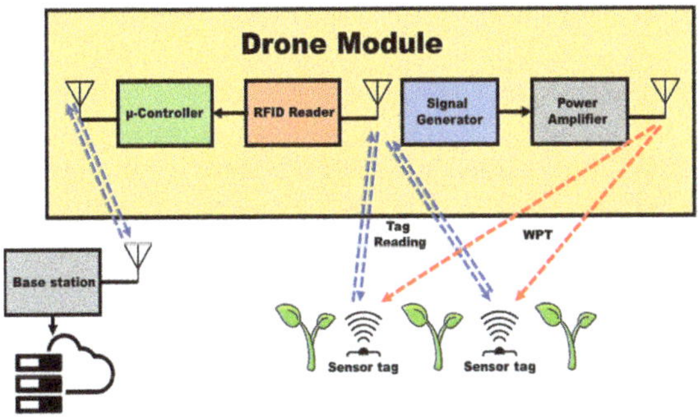

Figure 15. Communication module consisting of an RFID reader and WPT transmitter.

Based on the power consumption data in Table 3, which summarize the power consumption of the individual sensing components, the entire sensing module consumes around 0.5 mW that can be easily provided by the PMU storage capacitor, which can store up to 28.8 mJ of energy. This power can be provided for sufficient time for tens of read cycles from the RFID reader. The charging time depends upon the receiver's distance from the transmitter. As per the measured data extracted from Table 2, if the receiver is placed 1.6–1.87 m away from the transmitter, it will take approximately one minute for the PMU to charge the capacitor and be in a position to provide 0.5 mW to the sensing module for its operation. Similarly, if the receiver is placed 2–2.5 m far from the WPT transmitter, it will need approximately 2 min for the PMU to get charged and be able to provide 0.5 mW to the sensing module. Assuming a fully charged capacitor (28.8 mJ) and considering that a read cycle from any RFID reader takes less than 1 s, there is sufficient power for more than 60 read cycles before the EH system needs to be charged again.

Table 3. Power consumption of the sensing module.

Components	Power Consumption	
	Consumed Power	Unit
RFID IC ROCKY100	12.6	μW
MCU MSP430FR2355	468.6	μW
Humidity and Temperature Sensor SHT40	1.6	μW

For the conducted experiment, which took place in laboratory environment in order to control the orientation of the transmitter and receiver antennas, the UHF Fonkan reader used a transmission (TX) power of 23 dBm, while the gain of the transmitter antenna was 2 dBi. Under these conditions, the maximum read range of the passive tag was up to 2 m, and in the semi-passive or WPT-enabled mode, the best-case scenario read range was almost 5 m. When a power amplifier was added and a higher gain (7 dBi) antenna was used as the transmitter antenna, the transmitter power rose to 33 dBm, the read range of the passive UHF tag with the ROCKY100 RFID IC was tested up to 11 m, and for the semi-passive tag with the Energy Harvesting feature that was described in Section 3.2, the read range was 25 m (under optimal antenna orientation and polarization conditions).

The RFID tag read range was enhanced by operating the tag in semi-passive mode when the required DC power was provided from the RF Energy Harvesting module that was integrated with the humidity sensors. For the on-field testing, a customized UAV-carried reader and the customized sensors were used. Each sensing module consumes approximately 0.5 mW of power. This power can be provided for sufficient time from the energy that is stored in the storage capacitor (~28.8 mJ) assuming that the high voltage terminal of the PMU is used. The point-to-point communication between two microcontrollers (one deployed on the flying UAV and another at the base station) is effective up to 50 m using an omni-directional antenna at the base station. The maximum reading distance of the UAV reader depends on the EIRP. For the maximum transmitted power at +33 dBm and transmitter antenna gain at 7 dBi, the maximum reading range was 25 m for the semi-passive tag and 11 m for the tag operating in the passive mode. The RFID reader can simultaneously read more than 20 sensors scattered in an open field. The flying time of the UAV depends on the external load and the size of the battery.

The on-field testing demonstrated the feasibility of WPT from a UAV to wireless sensors with the Energy Harvesting feature. The conducted experiment demonstrated the effectiveness and the usefulness of using UAV-carried readers to collect data from battery-less "green" agriculture sensors.

4.3. UAV Readers with Beam-Forming Capability

The experiments indicated that using high-gain antennas can significantly increase the charging and the reading distances and can also support the simultaneous charging of several wireless sensor platforms with the Energy Harvesting feature. This observation led to the design of a UAV-carried beam-forming array which can direct the maximum gain of the radiation beam in five different pre-determined directions along two perpendicular planes [22]. The concept idea and a prototype conformal cross-shaped array designed for the ISM band at 2.4 GHz are presented in Figure 16. The schematic in Figure 16a demonstrates the concept of using beam-forming in order to increase the gain in the desired directions and eventually the communication (or power transfer) range between the UAV-carried transmitter and the ground-based sensors. Figure 16b shows the insertion loss of the conformal linear arrays when the maximum gain is directed in one of the three directions (0° and +/−37°). The three directions are achieved when the phase shifter causes a phase difference on the radiation elements (0° or + 120° or + 120°) and the plots for the two symmetric cases (+/−120°) coincide. In the inset, the prototype of the implemented orthogonal 3 × 1 linear array is presented.

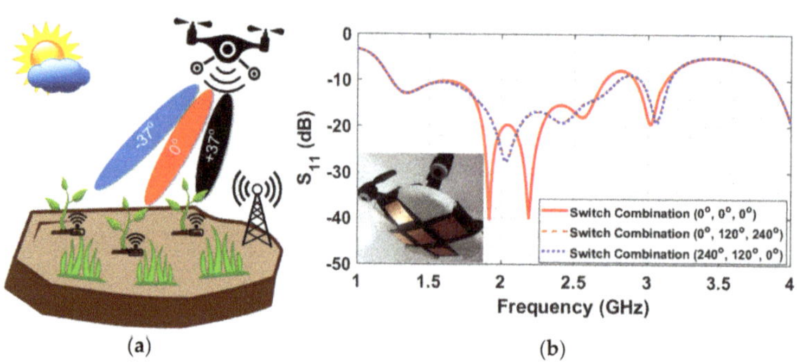

Figure 16. Beam-forming concept, and the insertion loss of the prototype cross-shaped array at 2.4 GHz.

For the prototype implementation, the ISM band was preferred because the two 3 × 1 linear arrays of patch radiators would be much bigger if the UHF band was used instead. In addition, the use of a beam-forming array is more useful for radiation in the ISM band because of the higher free-space losses that ISM frequencies suffer compared to the UHF band. Beam-forming is supported using a reconfigurable phase shifter in conjunction with a power combiner. For the three radiating elements in each of the two 3 × 1 arrays, the power ratios at 1:2:1 and the phase difference of the three adjacent driving currents can be either 0° or +120° or −120°. The power ratio is accomplished using the power divider presented in Figure 17a, which is a combination of three 3 dB Wilkinson dividers [22] arranged so they can have one input port and three output ports. The central port is directed on the central (common) patch element. The 1:2:1 power divider is followed by a three-stage phase shifter. The phase shifter is implemented using two different electric paths per branch, which are selected in combination with using three pairs of single-pole double-through (SPDT) switches [22]. The phase shifter schematic in Figure 15 shows the case when the three patch radiators are fed with 120° phase difference, causing the directivity of the 3 × 1 elements array to form a −37° angle. The set of six SPDT switches can be used in three combinations, and thus the main beam can be directed underneath the UAV to form a 37° angle with the vertical axis, either to the left or to the right. The maximum gain for the 3 × 1 array is 9 dBi and, under optimum orientation conditions, can quadruple the reading range between the UAV-carried reader and a ground-based sensor. In order to switch the feeding from one 3 × 1 array to the other, another three SPDT switches are used, as can be seen in Figure 17b. A switch is needed for the central patch because the feeding point along the patch is different in the two cases so as to maintain the matching. The patch arrays were implemented on a conformal substrate that was formed on the side of the 30 cm radius cylinder, and a Styrofoam supporting structure was implemented for maintaining the aforementioned shape.

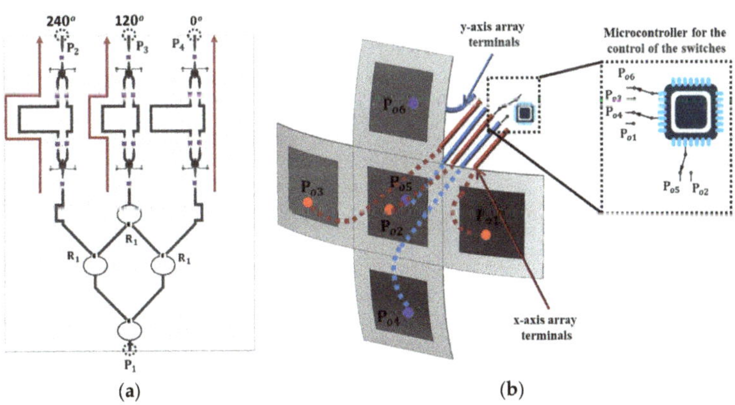

Figure 17. (**a**) Feeding Network consists of 1:2:1 power divider followed by a three-stage phase shifter (**b**) switching between the two orthogonal 3 × 1 linear arrays using SPDT switches controlled with a Microcontroller.

5. Conclusions

The manuscript provides a comprehensive report of the research findings of the research project ICARUS [23], which investigated the use of additive manufacturing, energy harvesting, and the use of UAVs as carriers of Wireless Power Transfer transmitters and their implementations as part of Precision Agriculture applications. 3D printing and inkjet printing were demonstrated for the implementation of passive humidity sensors. Based on the accuracy and repeatability limitations, the use of commercially available humidity and temperature IC sensors was considered. The IC sensors were integrated on sensor platforms that were using emerging harvesting circuits. Both UHF and ISM bands were utilized for communication and the energy harvesting module. It was demonstrated that a semi-passive RFID-based humidity sensor could quadruple its read range (+6 dB) when it was powered using harvested RF energy collected with the Energy Harvesting system integrated in the sensor's platform.

Finally, an experimental setup was described which used a UAV-carried WPT transmitter to charge a 4.7 mF capacitor used to store the harvested energy from a ground-based sensor platform. The efficient and effective charging of the customized Energy Harvesting system was demonstrated for distances up to several meters. Wireless charging was demonstrated with uni-directional antennas on the transmitter and the receiver for a distance up to 5 m, and, under optimal orientation and polarization conditions, it could be supported for a distance up to 18 m. The harvested energy was used to power the RFID IC that was able to increase its read range from 9 m to 25 m, proving that UAVs can be used to collect data and power RFID-based ground sensors suitable for Precision Agriculture applications.

Author Contributions: Conceptualization, L.V. and S.N.; methodology, A.N. and D.C.; validation, A.N., D.C. and L.V.; formal analysis, A.N. and D.C.; investigation, L.V.; resources, L.V. and A.N.; data curation, A.N.; writing—original draft preparation, L.V. and S.N.; writing—review and editing, S.N. and P.V.; visualization, A.N.; supervision, P.V.; project administration, S.N.; funding acquisition, S.N. All authors have read and agreed to the published version of the manuscript.

Funding: This research was funded by Cyprus Research and Innovation Foundation under the Project SMALL SCALE INFRASTRUCTURES/1222/0087 LARISSA-6G that is implemented under the Cohesion Policy Funds "THALEIA 2021-2027" with EU co-funding. This work was also supported in part by the European Commission Horizon 2020 programme via the CompSafeNano (Grant Agreement No. 101008099) project.

Institutional Review Board Statement: Not applicable.

Informed Consent Statement: Not applicable.

Data Availability Statement: The data presented in this study are available on request from the corresponding author.

Conflicts of Interest: David Chatzichristodoulou is the employee of the company RF and Microwave Solutions Ltd. The remaining authors declare that the research was conducted in the absence of any commercial or financial relationships that could be construed as a potential conflict of interest.

References

1. Singh, V. Advances in Precision Agriculture Technologies for Sustainable Crop Production. *J. Sci. Res. Rep.* **2024**, *30*, 61–71. [CrossRef]
2. Misra, N.N.; Dixit, Y.; Al-Mallahi, A.; Bhullar, M.S.; Upadhyay, R.; Martynenko, A. IoT, Big Data, and Artificial Intelligence in Agriculture and Food Industry. *IEEE Internet Things J.* **2022**, *9*, 6305–6324. [CrossRef]
3. Tagarakis, A.C.; Kateris, D.; Berruto, R.; Bochtis, D. Low-Cost Wireless Sensing System for Precision Agriculture Applications in Orchards. *Appl. Sci.* **2021**, *11*, 5858. [CrossRef]
4. Bošković, M.V.; Šljukić, B.; Vasiljević Radović, D.; Radulović, K.; Rašljić Rafajilović, M.; Frantlović, M.; Sarajlić, M. Full-Self-Powered Humidity Sensor Based on Electrochemical Aluminum–Water Reaction. *Sensors* **2021**, *21*, 3486. [CrossRef] [PubMed]
5. Kim, J.; Cho, J.-H.; Lee, H.-M.; Hong, S.-M. Capacitive Humidity Sensor Based on Carbon Black/Polyimide Composites. *Sensors* **2021**, *21*, 1974. [CrossRef] [PubMed]
6. Martins, J.E.M.P. Physical Analysis of a Waterproof Temperature Sensor Responsiveness for Agricultural Applications. *J. Agric. Inform.* **2023**, *13*. [CrossRef]
7. Song, C.; Liu, P.; Liu, X.; Liu, L.; Yu, Y. Configuration Optimization of Temperature–Humidity Sensors Based on Weighted Hilbert–Schmidt Independence Criterion in Chinese Solar Greenhouses. *Agriculture* **2024**, *14*, 311. [CrossRef]
8. Xu, J.; Gu, B.; Tian, G. Review of agricultural IoT technology. *Artif. Intell. Agric.* **2022**, *6*, 10–22. [CrossRef]
9. Chai, K.H.; Azizi, M.M.F.; Romeli, S.; Lau, H.Y. A Deployment of IoT sensors in the revolution of agriculture: IoT sensors in agriculture. *RA J. Appl. Res.* **2023**, *9*, 306–330. [CrossRef]
10. Bhandari, S.; Krishnanand; Singh, A.; Taufik, M. 3D printing methods and materials for sensor fabrication. *Mater. Today Proc.* 2023; ISSN 2214-7853. [CrossRef]
11. Gopalakrishnan, S.; Waimin, J.; Zareei, A.; Sedaghat, S.; Raghunathan, N.; Shakouri, A.; Rahimi, R. A biodegradable chipless sensor for wireless subsoil health monitoring. *Sci Rep* **2022**, *12*, 8011. [CrossRef] [PubMed]

12. Hassan, M.S.; Zaman, S.; Dantzler, J.Z.R.; Leyva, D.H.; Mahmud, M.S.; Ramirez, J.M.; Gomez, S.G.; Lin, Y. 3D Printed Integrated Sensors: From Fabrication to Applications—A Review. *Nanomaterials* **2023**, *13*, 3148. [CrossRef] [PubMed]
13. Bougas, I.D.; Papadopoulou, M.S.; Boursianis, A.D.; Kokkinidis, K.; Goudos, S.K. State-of-the-Art Techniques in RF Energy Harvesting Circuits. *Telecom* **2021**, *2*, 369–389. [CrossRef]
14. Dang, T.H.; Tran, V.T.; Chung, W.Y. B2EH: Batteryless BLE Sensor Network Using RF Energy Harvesting. In Proceedings of the IEEE Applied Sensing Conference (APSCON), Bengaluru, India, 23–25 January 2023; pp. 1–4.
15. Grosinger, J.; Görtschacher, L.; Bösch, W. Sensor add-on for batteryless UHF RFID tags enabling a low cost IoT infrastructure. In Proceedings of the IEEE MTT-S International Microwave Symposium (IMS), San Francisco, CA, USA, 22–27 May 2016; pp. 1–4.
16. Ayir, N.; Riihonen, T. Impact of Software-Defined Radio Transmitter on the Efficiency of RF Wireless Power Transfer. In Proceedings of the IEEE Wireless Power Transfer Conference (WPTC), Seoul, Republic of Korea, 15–19 November 2020; pp. 83–86.
17. Chien, W.C.; Hassan, M.M.; Alsanad, A.; Fortino, G. UAV–Assisted Joint Wireless Power Transfer and Data Collection Mechanism for Sustainable Precision Agriculture in 5G. *IEEE Micro* **2022**, *42*, 25–32. [CrossRef]
18. Liu, Z.; Li, J. Application of Unmanned Aerial Vehicles in Precision Agriculture. *Agriculture* **2023**, *13*, 1375. [CrossRef]
19. Chatzichristodoulou, D.; Quddious, A.; Vassiliou, L.; Shoaib, N.; Vryonides, P.; Nikolaou, S. Additive manufacturing techniques for 5G IoT antennas and sensors. In Proceedings of the IEEE 15th European Conference on Antennas and Propagation (EuCAP), Dusseldorf, Germany, 22–26 March 2021; pp. 1–5.
20. Chatzichristodoulou, D.; Quddious, A.; Paolini, G.; Masotti, D.; Costanzo, A.; Vryonides, P.; Nikolaou, S. UHF and ISM Band MIMO Antenna for an Energy-Autonomous SWIPT system. In Proceedings of the 2022 IEEE International Symposium on Antennas and Propagation and USNC-URSI Radio Science Meeting (AP-S/URSI), Denver, CO, USA, 10–15 July 2022; pp. 191–192.
21. Nadeem, A.; Chatzichristodoulou, D.; Vassiliou, L.; Vryonides, P.; Nikolaou, S. Energy Harvesting Supported Humidity Sensor for Agriculture Applications. In Proceedings of the Sensors and Electronic Instrumentation Advances (SEIA), Corfu, Greece, 21–23 September 2022; p. 161.
22. Nadeem, A.; Chatzichristodoulou, D.; Kiani, H.; Quddious, A.; Vassiliou, L.; Shoaib, N.; Vryonides, P.; Nikolaou, S. Transmitter Considerations for 5G Wireless Powered Sensor Networks. In Proceedings of the 16th European Conference on Antennas and Propagation (EuCAP), Madrid, Spain, 27 March–1 April 2022; pp. 1–5.
23. Frederick. Available online: https://icarus.frederick.ac.cy/ (accessed on 15 May 2024).

sensors

MDPI

Article

The Rise of Passive RFID RTLS Solutions in Industry 5.0

Ygal Bendavid [1,*], Samad Rostampour [2], Yacine Berrabah [1], Nasour Bagheri [3] and Masoumeh Safkhani [4]

[1] Department of Analytics Operations and Technology, School of Management, Université du Québec à Montréal (UQAM), Montreal, QC H2X 3X2, Canada
[2] Department of Computer Science, Vanier College, Montreal, QC H4L 3X9, Canada; rostamps@vaniercollege.qc.ca
[3] CPS[2] Lab, Electrical Engineering Department, Shahid Rajaee Teacher Training University, Tehran 16788-15811, Iran; nbagheri@sru.ac.ir
[4] Computer Engineering Department, Shahid Rajaee Teacher Training University (SRTTU), Tehran 16788-15811, Iran; safkhani@sru.ac.ir
* Correspondence: bendavid.ygal@uqam.ca; Tel.: +1-514-987-3000

Abstract: In today's competitive landscape, manufacturing companies must embrace digital transformation. This study asserts that integrating Internet of Things (IoT) technologies for the deployment of real-time location systems (RTLS) is crucial for better monitoring of critical assets. Despite the challenge of selecting the right technology for specific needs from a wide range of indoor RTLS options, this study provides a solution to assist manufacturing companies in exploring and implementing IoT technologies for their RTLS needs. The current academic literature has not adequately addressed this industrial reality. This paper assesses the potential of Passive UHF RFID-RTLS in Industry 5.0, addressing the confusion caused by the emergence of new 'passive' RFID solutions that compete with established 'active' solutions. Our research aims to clarify the real-world performance of passive RTLS solutions and propose an updated classification of RTLS systems in the academic literature. We have thoroughly reviewed both the academic and industry literature to remain up to date with the latest market advancements. Passive UHF RFID has been proven to be a valuable addition to the RTLS domain, capable of addressing certain challenges. This has been demonstrated through the successful implementation in two industrial sites, each with different types of tagged objects.

Keywords: IoT; passive UHF RFID; RTLS; real-time location systems; IPS; indoor positioning systems; industry 5.0; digital transformation

1. Introduction

Beginning in the 2000s, the North American manufacturing sector gradually lost significance. This decline in manufacturing activity was driven by various factors, including growth in service industries, alongside strong competition from Asian countries, which leverage cheaper labor and extensive production expertise. In addition to these combined effects, there was a lack of investment in technology by manufacturing companies. To compound matters, the COVID-19 pandemic dealt a severe blow to the sector in 2020.

To address this sectoral weakness, stimulate manufacturing, enhance productivity, and support competitiveness, governments worldwide proposed digital economy action plans. These plans underscored the crucial need for modernization and adoption of Industry 4.0/5.0 concepts [1], often referred to as Industry X.0, as coined by Schaeffer in 2017 [2].

While strategically, digital transformation forces companies to reassess their business models and management approaches, practically, many companies start this journey by trying to solve operational problems [3]. Indeed, an initial focused project often serves as a gateway for companies motivated by practical business concerns, rather than a shift in mindset towards services, value, and outcomes. As Ferreira et al. [4] point out, SMEs are increasingly interested in moving towards Industry 4.0, whether driven by internal motivation or by pressure (from customers and/or large companies), especially if the project aims at cost reduction and short-term benefits (e.g., flexibility, efficiency). This is the case for many companies that are experiencing productivity losses due to the lack of real-time visibility into the location of their critical assets and logistics units.

While the challenge of selecting the most suitable technology design for a specific use case is common to many companies, the market offers a portfolio of indoor Real Time Location Systems (RTLS) for tracking and tracing people, products, and equipment. These solutions predominantly rely on competing geolocation technologies, with "active" transponders (tags that require batteries to power the asset tracking devices) being the dominant choice. However, for managing dormant equipment, such as items used sporadically and then stored for extended periods, solutions requiring tag battery maintenance or replacement pose drawbacks. Conversely, the emergence of new RTLS based on "passive" Radio Frequency Identification (RFID) technologies is disrupting the market, offering tags that do not rely on batteries for power [5].

As a result, several companies have taken advantage of technological innovations in the field of passive RFID and developed RTLS readers accordingly. Impinj (Seattle, WA, USA) led the way in 2014 with its passive RTLS reader, the xArray reader (2D x, y), followed by RF Controls in 2017 and Zebra in 2020 (3D x, y, z). Despite notable performance enhancements over the years, the adoption of these systems remains limited. Moreover, the academic literature on passive RTLS is sparse, often incomplete, or outdated. Recent literature reviews indicate a predominant focus on passive RFID technologies for real-time tracking and monitoring applications in logistics and supply chain management [6,7], leaving passive RTLS with minimal or no coverage. Other studies in the field of RTLS are primarily conducted by solution vendors, raising concerns about result neutrality.

This gap in the literature raises several questions about the most appropriate design for asset tracking (physical and software infrastructure) and the actual performance versus that advertised by vendors of alternative RTLS solutions, including those based on passive RFID technologies. For instance,

- Can passive RFID be considered a serious competitor to established RTLS technologies in the market?
- What key features should be used for comparing indoor passive RTLS technologies?
- What is the realistic performance of passive RFID RTLS technologies?

Objective of the Research

This research project aims to (a) propose an up-to-date typology of RTLS and (b) clarify the realistic performance of passive RTLS solutions. To achieve the first objective, we review both the academic and industry literature. To achieve the second objective, following a design science approach, we develop and test an RTLS prototype based on passive RFID technologies in two different industrial environments.

In Section 2, we explore the central role of IoT technologies supporting innovative applications in Industry 5.0, particularly those related to RTLS solutions. We conduct a review of the academic and professional literature to identify the knowledge gap related to passive RTLS.

In Section 3, we present an updated, synthetic comparative analysis of RTLS technologies to reflect the latest industry trends.

In Section 4, we introduce a case study to anchor our research problem, aiming to validate the performance of emerging passive RTLS systems. We follow a design science research methodology to develop our artifact (the prototype solution).

Finally, in Section 5, we conclude and open the discussion on security issues related to emerging passive RTLS.

2. Literature Review

2.1. Industry 4.0 to 5.0

As Garms [8] points out in a McKinsey report that draws insights from companies adopting Industry 4.0, "Succeeding in Industry 4.0 requires more than just using new IT tools; it demands a comprehensive tech-driven overhaul of operations". This approach to digital transformation involves automating processes and integrating new technologies throughout the company's value chain, linking different systems within the plant, and synchronizing them. Industry 4.0 aims to use digital technologies to enhance the efficiency of value chain activities [9], respond to the increasingly personalized needs of customers, and adapt more quickly to market changes. Building on this concept, Industry 5.0, as proposed by the European Commission [10] retains the technological aspect of 4.0 but emphasizes three pillars: (a) business agility and resilience supported by advanced technologies, (b) prioritizing humans in innovation and competitiveness by focusing on skills development, and (c) embracing environmental sustainability and considering the product life cycle. A recent systematic literature review by Valette et al. [11] underscores the rising importance of Cyber-Physical Systems (CPS) and IoT concepts in industrial research, with a strong emphasis on human-centricity. They advocate for viewing future industrial systems as complex socio-technical entities, where human and industrial assets are seen as interconnected. Galdysz et al. [12] conduct a systematic literature review on the transition to Industry 5.0 and find that the integration of the human factor is often overlooked. They suggest exploring the "grey" literature to enhance our understanding and recommend practical studies to implement technical concepts. To address these insights, our paper will (a) incorporate real-world deployments of RTLS in manufacturing contexts and (b) present a case study in the paper's final section to provide a practical grounding. This approach aligns with Leng et al.'s [1] observation that few papers discuss the implementation journey of Industry 5.0.

2.2. IoT in the Industry X.0 Technology Portfolio

The Internet of Things (IoT) plays a vital role in enhancing business flexibility and resilience, serving as a key component of Industry 5.0 [1]. It creates a technological environment where every physical object, whether living or not, can communicate automatically and in real-time with its surroundings, managing transactions efficiently [13]. Each connected object becomes a cyber-physical device, contributing to a smarter ecosystem. De Paula Ferreira [14] identifies "smart products", which are uniquely identifiable and constantly tracked items, as crucial for implementing digital transformations in Industry X.0.

Research on this topic has surged over the past two decades across various industries, with a particular focus on Industrial IoT (IIoT) systems [15], which support operational processes [8,16]. The interaction facilitated by IoT technologies between physical and virtual realms has led to the emergence of digital twins, where physical entities have real-time updated digital replicas [17,18] throughout their lifecycle.

Figure 1 shows the infrastructure of an IoT solution derived from various models proposed in the literature (adapted from [13]).

This model helps us to grasp that implementing an IoT solution involves the integration of multiple evolving technology layers based on various protocols and standards for (a) object identification and connectivity, (b) real-time data capture by readers, (c) wireless communication networks, (d) local or cloud-based data storage platforms, and (e) software platforms for visualization, analysis, and integration with back-end systems. AI (Artificial Intelligence) advancements over the past decade have integrated vision recognition solutions into this portfolio, where objects serve as identifiers recognized by algorithms. AI also enhances communication network performance and database management and primarily facilitates extensive data analysis and decision-making in various management systems.

Our research focuses on RTLS solutions within the (i) identification and (ii) data capture layers of an IoT infrastructure. These location-based functions are vital in the context of "smart factories", [19] and IoT technology innovations have made implementing such systems more feasible, as seen in various offerings by technology providers. In a systematic mapping study, Haibi et al. [20] highlight a significant and ongoing increase in RFID research activities worldwide, particularly in supply chain and logistics management. However, only a small percentage of these studies address localization techniques. Despite their significance in the IoT realm, the academic literature shows limited interest in the design and development of localization solutions based on passive technologies [19]. Some studies [21] propose such solutions but tend to rely on multiple fixed readers rather than passive RTLS readers. Specifically, while RFID technology garners significant interest [6], deployment in logistics initiatives primarily focuses on tracking (41.67 % of the results) and monitoring (31.25%) applications [7].

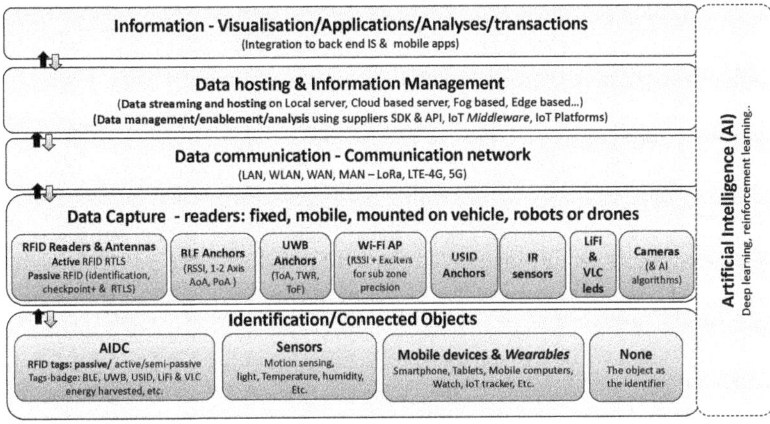

Figure 1. IoT multi-layer infrastructure: emphasis on RTLS.

2.3. *Summary Analysis of RTLS in the Academic Literature*

It is important to note that in the literature, Indoor Positioning Systems (IPS) typically refer to location-based services (LBS) on mobile devices, while RTLS generally focuses on locating objects using tags. Both terms are linked with indoor location-based services (ILBS), which emphasize both location and using processed data for decision making.

From academic and professional perspectives, various comparative analyses aim to assess the performance of RTLS/IPS, but very few consider passive RTLS as an option.

For example, in a recent survey of Indoor Location Technologies, Techniques, and Applications in Industry, Hayward et al. [22] summarize commercially available technologies. They use various criteria like cost, privacy/security, power consumption, coverage, scalability, accuracy, interference, and signal requirements. However, their analysis of BLE or RFID technologies is quite simplistic. They fail to break down the technologies and techniques to accurately represent recent market developments. Specifically, for passive UHF RFID technologies, they suggest coverage ranging from 10 m^2 to over 1000 m^2 with an accuracy of 1 m to 5 m.

Farsani et al. [23] propose an interesting survey on Indoor Positioning Systems for IoT-Based Applications. The authors discuss IPS classification, IPS algorithm classifications, and enabling communication technologies and discuss common metrics in IPS comparisons: accuracy (closeness of the measured or estimated position to the real one) and precision (closeness of a quantity in repeated measurements over time). Unfortunately, they do not include availability–accessibility of a service, cost, coverage, energy efficiency, latency, scalability, and robustness. Unfortunately, in their analysis, they completely omit the latest developments in passive RFID technologies.

Zafari et al. [24] addressed the need for an updated survey paper incorporating recent accurate and reliable localization systems. In their work "Emerging IoT Technologies-Based Localization", they discuss Low Power Wide Area Networks (LPWANs) like LoRA, which require integration with other technologies for localization. However, their analysis of Passive RFIDs is oversimplified, indicating a "limited communication range (1–2 m)" and relying on prior research with fixed RFID readers.

Halawa et al. [25] narrow down RTLS technology choices to UWB, RFID, WiFi, and Vision. They specifically integrate vision-based object recognition systems; while they consider the "3D location" criterion achievable with various solutions, their assessment of RFID technologies is broad, grouping all types into one category.

The authors of [26] propose a comparative study on "Seamless Asset Location and Tracking Technologies". They present an evaluation matrix comparing tracking approaches for both indoor and outdoor technologies. Although they differentiate between active and passive RFID technologies, their analysis overlooks the potential of the latest passive RFID technologies for localization.

Asaad and Maghdid [27] explored various metrics to compare localization solutions within the IoT landscape, including accuracy, precision, cost, scalability, latency, security, reliability, and system complexity. They rightly highlighted the complexity of indoor RTLS deployment compared to open-air settings. The authors mention that passive RFID technologies "are gaining popularity for localization due to their low cost, ease of installation, and maintenance", but they do not elaborate on this discussion.

The authors of [28] examined RTLS applications in manufacturing to offer a comprehensive framework for positioning available RTLS solutions. Unfortunately, their discussion excluded passive RTLS from consideration.

Comparative analyses of RTLS technologies often fail to distinguish between evolving technologies. For instance, Bluetooth technologies now include BLE AoA (angle of arrival), BLE TDOA (time difference of arrival), and BLE RSSI (received signal strength indicator). Similarly, studies often lump together checkpoint RFID solutions and RTLS solutions equipped with multi-beam antennas.

The academic literature on passive RTLS is scarce, incomplete, or outdated, with most studies provided by solution vendors, raising concerns about the neutrality of the findings. The comparison presented in Table 1 outlines the characteristics of the referenced papers. Within the contribution column, each paper's unique advantage is highlighted. The active RFID column denotes whether each paper addresses this technology. In the passive RFID column, the discussion of RTLS Passive UHF RFID specifics by each paper is indicated. Additional information is provided in the final column.

Table 1. Comparison of academic related works.

Ref.	Contribution	Active RFID	Passive RFID	Additional Comments
[22]	Comparative analysis of indoor location-based services (ILBS): GNSS, Inertial System (MEMS), Wi-Fi, Bluetooth LE, WSN (Zigbee), UHF RFID (Passive), UHF RFID (Active), Ultra-Wideband, Acoustic (Ultrasound), Vision	YES	No	No consideration of BLE or passive RFID development
[23]	Comparative analysis of indoor positioning system (IPS): BLE5, WIFI, UWB, RFID Zigbee, VLC Acoustic, Ultrasonic, Vision, LoRa Sigfox and Cellular Comparative positioning techniques (RSSI, ToA, TDoA, RTT, etc.)	YES	No	No consideration of passive RFID technologies as an IPS candidate

Table 1. *Cont.*

Ref.	Contribution	Active RFID	Passive RFID	Additional Comments
[24]	Comparative analysis of Wireless technologies for Localization (IPS): IEEE 802.11 n/ac/ad/ah UWB, Acoustic, RFID' Bluetooth, Ultrasound, Visible light, SigFox, LoRa, Weightless; Comparative localization techniques (RSSI, CSI, AoA, ToF, TDoA, RToF, PoA, Fingerprint)	YES	No	Active RFID "generic" No Consideration of passive RFID technologies as an IPS candidate
[25]	Comparative discussion on RTLS: UWB, RFID systems, vision systems, and Wi-Fi technology. RTLS technology evaluation	YES	No	Emphasis on UWB Technologies for RTLS. RFID is considered a tracking technology
[26]	Comparative Analysis among Tracking Approaches (RFID Active, RFID passive, BLE, UWB, Wi-Fi, LoRaWan, Cellular, GPS, Image)	YES	No	Segmentation of Active and passive RFID - where passive RFID is identified as short-range tracking technology vs RTLS
[27]	Comparative analysis of Localization Technologies and techniques GNSS (TOA, TDOA), Optical (TOA, AOA. TDOA, RSSI, Cell-Id), Wi-Fi (RSSI, RTT, TOA, TDOA, AOA, AP-Id), UWB(TOA, TDOA, RSSI, AOA), RFID (AP-ID, RSSI), Bluetooth (AP-ID, RSSI, TOA), FM, Cellular, Sound	YES	No	Point out passive RTLS developments but don't discuss it
[28]	Qualitative comparison of RTLS-related communication protocols (UWB, Wi-Fi, BLEGPS, RFID, 5G)	YES	No	Consider passive RFID as a close-range technology

2.4. *Summary Analysis of RTLS in the Professional Literature*

The Established Active RTLS Technologies

In both the academic and professional literature, there is a wealth of information about active-based RTLS technologies, covering both current and emerging options. For instance, [29] from the *Sensors* journal discusses Wi-Fi as a promising option for indoor positioning, highlighting important positioning schemes and addressing challenges like the multi-path effect and data privacy. Another study by [30] explores the potential and limitations of BLE technologies, focusing on using RSSI measurement for indoor positioning and suggesting BLE AoA for future research. Additionally, [31] compares different wireless communication-based technologies for IPSs, particularly focusing on UWB-IPS for IIoT and emphasizing the characteristics of UWB technologies, such as large bandwidth and high data rate, crucial for precise indoor localization accuracy. However, they overlook passive RTLS in their analysis. On a different note, [32] delves into the advancements in IPS via visible lights, presenting it as a forthcoming low-cost solution for indoor localization.

In the professional literature, numerous documented cases showcase the deployment of these technologies across various business fields, demonstrating their effectiveness and versatility. There are also several IPS/RTLS comparative analyses proposed in the form of white papers, where each vendor promotes its technology as the best. For example, vendors like BlueIoT and Quuppa advocate for BLE AoA technology, emphasizing its high location accuracy, low power consumption, and compatibility with numerous BLE devices. On the other hand, UWB vendors like Ubisense and Sewio highlight the high accuracy and longer range of UWB technology. However, these reports often oversimplify the comparative analysis and neglect passive UHF RFID technologies.

For a deeper market analysis, research and consulting service providers like Gartner offer comparative analyses of IPS, although they mainly focus on active technologies. Figure 2 presents an adapted version of Gartner's "2022 and 2023 Magic Quadrant for Indoor Location Services" (adapted from [33,34]). The leading vendors are positioned in a two-axis map of (a) completeness of vision and (b) ability to execute. We modified this matrix to highlight the variety of competing technologies and positioning techniques in the market.

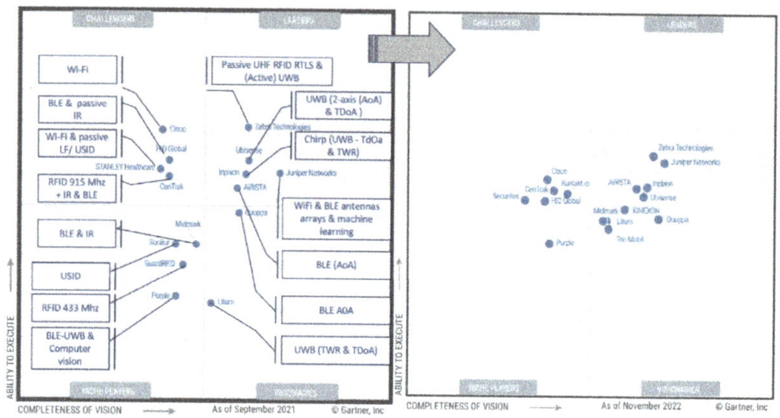

Figure 2. Adapted version of Gartner Magic Quadrant for indoor location services—2022 and 2023.

2.5. *Actual and Emerging Indoor RTLS Options*

One common approach is WiFi-based RTLS, utilizing antennae, proximity detection devices, and efficient algorithms for accurate location. Well-known vendors in this domain include Cisco, Aruba, and Securitas (formerly Stanley Healthcare).

Bluetooth Low Energy (BLE) has also made strides in RTLS. Many existing BLE-based solutions rely on received signal strength (RSSI), limiting accuracy to 4–5 m. Notably, vendors like HID Global and Minew operate in this space. Some, like Arista or Quuppa, employ BLE/AoA for more precise localization accuracy, reaching a few centimeters.

Ultra-wideband (UWB) is widely used in RTLS and is less affected by signal interference and capable of penetrating various materials. With localization accuracy ranging from 10 to 25 cm, UWB utilizes triangulation algorithms based on Time of Arrival (TOA), Time Difference of Arrival (TDOA), and more recently, Time of Flight (TOF). Zebra, Sewio, Juniper, and Ubisense are established vendors in this sector.

Ultrasound identification technologies (USID) offer reliable indoor tracking with centimeter-level accuracy, boasting energy efficiency and easy deployment. However, the speed of sound variation due to environmental factors affects signal performance. Sonitor is a prominent market leader, now enabling the geolocation of smartphones.

Visible Light Communication VLC and Light Fidelity (LiFi) using light sensors to measure the position and direction of LED emitters, while providing high accuracy, line-of-sight between the LED and sensor is necessary. Companies like Lumentrace combine VLC-based localization with Bluetooth communication.

Active RFID technologies, operating in various frequency bands (6–8 GHz, 2.4 GHz, 433 Mhz, 915 Mhz), face limitations due to standardization issues and the need for additional infrastructure for localization. Hybrid active RFID tags, operating on different frequency bands, employ sensors like infrared, low-frequency, or ultrasonic for precise identification in specific areas. Centrack for instance uses active 915 Mhz tags also equipped with infrared (IR) sensors for precise location.

The portrait of the industry therefore presents multiple technological options; nonetheless, it is incomplete. The inclusion of passive RTLS in this portfolio of solutions is now a must. There is therefore a need to update and validate RTLS technology categories in the academic and professional literature reviews.

While there is also a wealth of the academic and professional literature regarding passive RFID technologies (battery-less tags activated by electromagnetic emitted by the antennae), most documented cases of these technologies in business fields, like those found in *RFID Journal*, involve fixed or portable readers. In the case of RTLS, which relies on passive RFID technologies, tag localization is now made possible by the use of omnidirectional (multi-beam) phased-array antennas. This technology was first introduced in 2008 by Mojix and developed over the following years, still struggling to establish itself in the market. The first passive RTLS cases emerged in the retail sector after Impinj introduced the Power-over-Ethernet (PoE) phased-array antenna xArray ceiling reader in late 2014 [35]. This reader is equipped with 52 beams with both horizontal and vertical polarizations covering up to 1500 ft^2 (140 m^2) within 1.5 m (5 feet) of an item's actual location.

In 2017, RF Controls introduced the CS-445B passive RTLS reader antenna, featuring an overhead bidirectional, steerable array. Revealed at the 13th Annual RFID Journal Live, it won the Best New Product award. The first generation of the CS-445B model was designed for ceiling heights of 10 to 30 feet, with a 45-foot read range and the ability to locate tagged items with an accuracy of about 18 inches [36]. In 2020, RF Controls launched the CS-490 antenna, reading tags up to 50 feet high and reaching a maximum distance of 90 feet, suitable for warehouses and manufacturing plants. This version offers 3D positioning (x, y, z) and higher accuracy when multiple antennas are used. In open spaces, mounted at 30 feet, it can cover a 3600 ft2 scan area. RTV Engineering and RF Controls deployed the technology in a manufacturing company to identify, locate, and track materials stacked seven bins high on metal racks facing each other across an aisle. The deployment of five RF Control antenna units, at a height of 32 feet, covered a 100-foot aisle with 10-inch accuracy [37]. Recently, Tego, an Industrial IoT (IIoT) solution provider, partnered with RF Controls to deploy a passive RFID-based RTLS for aerospace and defense companies. The technology, installed 50 feet above the floor, accurately locates, tracks, and encodes passive UHF RFID tags attached to forklifts, returnable transport items, and assets, with an accuracy of 1 to 3 feet [38].

In addition to enhancing reading distance and location accuracy, it is worth mentioning the distance-encoding feature, which allows pinpointing a specific tag's location for encoding, even from a distance of 500 feet [39]. Continuing in the same line of product development, Zebra Technologies, a leading RFID-IoT vendor, launched its ATR 7000 high-ceiling overhead reader in late 2019, enabling RTLS solutions with passive tags [40]. This reader works alongside their license-based RTLS software, CLAS (Configuration, Location Analytics Software), to visualize tag locations on a map. Zebra also introduced the ZBR4000 passive RFID tags, designed to work seamlessly with the ATR. One of the earliest deployments occurred at a US transportation company, enhancing real-time inventory tracking and facilitating warehouse cross-dock activities [41]. Zebra utilized this pilot project to enhance its "Motion Works" platform, providing visibility into operations and streamlining integration with existing warehouse management systems (WMS).

3. Redefining the Portfolio of RTLS Options

3.1. The Emergence of RTLS Based on Passive RFID

Based on previous analyses, we propose an updated portfolio of RTLS summarizing the different technologies used in RTLS solutions (see Figure 3).

This figure presents a current overview of the various technologies utilized for item identification and real-time tracking. The figure is segmented into four technological blocks. The first block represents the software infrastructure used to manage RFID-IoT data which is required whatever the choice of the hardware used to identify and locate items. Then, the next three blocs represent different types of hardware infrastructure; while these technologies often compete, they can also complement each other when integrated into a single solution, leveraging the strengths of each system. For instance, one technology might allow full zone coverage with low accuracy for detecting tagged items, while another technology would precisely identify the transit of tagged items between zones. These blocs are discussed in more detail below.

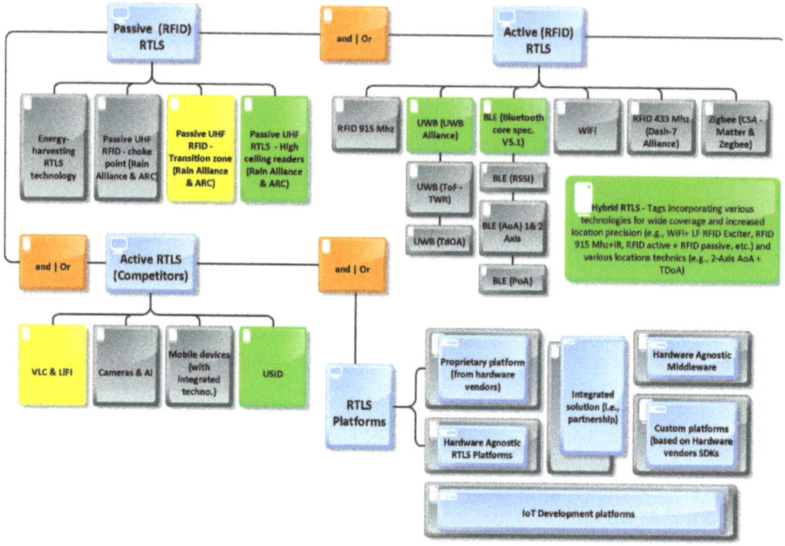

Figure 3. Updated indoor RTLS portfolio.

The first block "RTLS Platforms" concerns the software platforms used to manage RFID-IoT data gathered from the hardware infrastructure. In the case of active RFID solutions, there are several business models: (i) proprietary platforms, in which case the tag supplier also offers its own platform (e.g., Mobileview from Securitas, Quuppa Software suite or Simatic RTLS from Siemens), (ii) agnostic software platform providers whose solution enables the management of the data obtained, including businesses processes and workflow content, as well as dashboards and integration with management systems (e.g., Tego's Asset Intelligence Platform-AIP; Catamaran NextGen from Shipcom), and (iii) more simplistic middleware enabling basic but crucial functions such as the acquisition, the processing, and the transfer of the data obtained to back end systems (e.g., Pareto Anywhere from reelyActive or Clearstream from PTS Mobile). IoT platforms from well-known vendors can also be used to develop robust RTLS solutions (e.g., Microsoft Azure IoT, Amazon AWS).

The second block "Active (RFID) RTLS" represents established active RFID technologies (such as BLE, WIFI, and active RFID), each employing various techniques. For instance, BLE technologies utilize methods like the received signal strength indicator (RSSI) for localization accuracy within a few meters, BLE Angle of Arrival (AoA) for sub-meter accuracy, and BLE AoA on two axes (y, z) for even greater precision. Additionally, BLE phase-ranging technologies extend battery life. It is worth noting that the trend towards hybrid tags, which combine features like wide coverage detection with precise localization, often incorporating ultrasonic sensors or low-frequency receivers. In another scenario, a tag could be equipped with a BLE tag to calculate location and employ low-power wide-area (LoRa) gateways to forward data to a server.

The third block "Active RTLS (competitors)" consists of other battery-equipped technologies competing with active RFIDs, including VLC (e.g., Lumentrace), LiFi, and USID (e.g., Sonitor), as well as cameras with AI algorithms and increasingly advanced detection capabilities in mobile devices.

The fourth block "Passive (RFID) RTLS", of particular interest, encompasses the portfolio of passive RFID readers, which fall into four main families presented in the next section.

3.2. *Key Players and Technologies in the Field of Passive RTLS*

First, we find fixed readers used as checkpoint types (e.g., gates) that detect the presence of a tagged object when it passes within the radiation field of the reader's antennas. In order to determine the direction of the tag, antennas must be installed or external devices (e.g., motion detector, laser) are required.

Second, there are movement detection readers (e.g., directional gates). These readers are equipped with at least two antennas that detect (a) the presence of a tagged object and (b) the direction of passage of the tagged object (e.g., Impinj xSpan gateway, Kathrein ARU8500 reader, and Zebra ST5500 Transition RFID Reader)

Third, "real" passive RTLS readers include Impinj's xArray which offers the possibility of geolocating a tagged object with precise coordinates (x, y) in a ceiling of fewer than 15 feet. As discussed, more recently, other passive high ceiling RTLS readers equipped with Steerable Array Antenna from RF Controls (Smart Antennas) and Zebra (ATR 7000) which provide coordinates in 3D (x, y, z) are now redefining the landscape. A comparative matrix of such readers is shown in Table 2. In this table, various features are presented for comparing indoor passive RTLS technologies. For example, coverage indicates the area that antennas cover. Location accuracy indicates the level of precision and error in detecting the location of each object. The reading distance is the maximum readable distance from a tag to a reader. The location dimension shows the number of dimensions in each technology, which can be two-dimensional or three-dimensional.

Fourth, in parallel to the development of readers, one should note the development of energy harvesting tags. Indeed, while most passive RFID tags reflect back radio waves from the reader (responding), energy harvesting is a technique in which energy (from the reader) is gathered by the tag, stored briefly, and transmitted back to the reader–receiver (e.g., [42]). Uwinloc is an established provider of such solutions in the Industry 5.0 Context. Their solution combines (a) RFID UHF Exciters located near storage-detection areas (b), powering passive harvesting tags (c) which emit their unique ID back to the UWB receiver (d), connected to Wi-Fi or ethernet network (e), to send the data to a location engine processing unit.

With continued innovation and performance improvements, passive RTLS are then competing in a market traditionally reserved for active RTLS. The next logical step for researchers is therefore to verify the performance of such systems.

Table 2. Updated indoor passive RTLS.

	Impinj	**Zebra**	**RF Controls**
Passive RTLS Reader	xArray	ATR7000	CS-445B Series and CS-490
Reader & steerable phased array antenna ETSI-Lower Band 865–868 MHz ETSI-Upper Band 915–921 MHz were allowed			
Coverage (sq. ft)	1500 sq. ft	N/A	10,000 sq. ft
Location accuracy (ft)	1–1.5	2	1.5–3
Reading distance (ft)	30 ft. radius	45 ft. radius	45–75 ft. radius
Distance encoding	No	No	Yes
Deployment - Mounting height (ft)	15	12–18	10–30/20–50
Power source	PoE	PoE+	PoE+
Location-Dimension	2D	3D (with 2 readers)	3D (with 2 readers)
Platform\OS	Linux	CLAS\Linux	RFC-OS
Directional readers (Zone monitoring and directional readers)	xSpan Gateway (1000 sq. ft)	ST5500 Transition RFID Reader	None
Other vendors: Kathrein Solutions ARU 8500 Antenna Reader Unit			

4. A RTLS Case Study in Manufacturing

This case study takes place in the manufacturing industry where many companies have developed their core competencies around product development and manufacturing, paying less attention to the logistics aspect of their operations. Consequently, these companies often encounter logistical challenges because they lack insight into their ongoing operations and the whereabouts of their crucial mobile equipment. To address these issues, many companies turn to Real-Time Location Systems (RTLS). However, due to the rapid growth of various Internet of Things (IoT) technologies, architectures, and platforms in recent years, choosing, developing, and implementing an Industrial Internet of Things (IIoT) system has become exceedingly complex and time-consuming. It requires extensive technical expertise, which poses significant barriers to entry into the market, especially for small and medium-sized enterprises (SMEs) [43].

4.1. Defining the Motivation on the Case

Company A is a manufacturing company specializing in the production of food packaging and containers. The company's lack of real-time visibility into the location of hundreds of critical pieces of equipment (specifically, the molds used to make plastic containers) has a direct impact on operations planning, forcing the company to reschedule production.

4.1.1. Definition of the Business Problem

Because molds can be stored for months, even years, before being placed back on a press to produce the appropriate range, the problem is not immediately reported but arises during production planning. Typically, after each use, the mold is placed in a wooden crate and stored in the warehouse until the next use. Although the mold's location is recorded in the system (by scanning its bar code and location), the reliability of the information is sometimes compromised. For example, the physical location of these molds is sometimes not updated in the ERP system (molds moved without notification, stored in inappropriate locations, borrowed without notification, etc.). However, given the large number of molds and the size of the multi-zone, multi-row, multi-shelf warehouse, the search for a specific mold can take several hours to several days—sometimes without being found. As a result, the search for a mold wastes a lot of time, mobilizes employees in a non-value-added activity, causes production delays, leaves machines idle, and significantly reduces productivity. If a mold is not found, it also means re-manufacturing costs for a new mold.

4.1.2. Design Science Research Approach to Build the Case

In this research, we adopted the "Design Science" methodology used in information systems projects, which focuses on solving practical problems through the design and development of an artifact, i.e., our passive RTLS prototype. This method proposes several stages, as shown in Figure 4 [44]. Given the scope of the pilot project, we have limited our approach to the first four steps of the methodology, see Figure 5.

4.2. Definition of the Pilot Project

To address this issue and ensure the visibility of this critical equipment, a "passive" indoor positioning solution is being considered.

4.2.1. Objective of the Pilot Project

Given the defined context, the choice of such a solution must take into account several criteria: (i) require minimal maintenance of the tags since these molds can be stored for several months to several years before being reused; (ii) take into account the imposing structure of metal molds and their handling in the warehouse and the machines; (iii) ensure the detection of the molds everywhere in the areas of the warehouse equipped with 15 m high metal shelves; and (iv) be used in other possible use cases in the future, in particular for production monitoring.

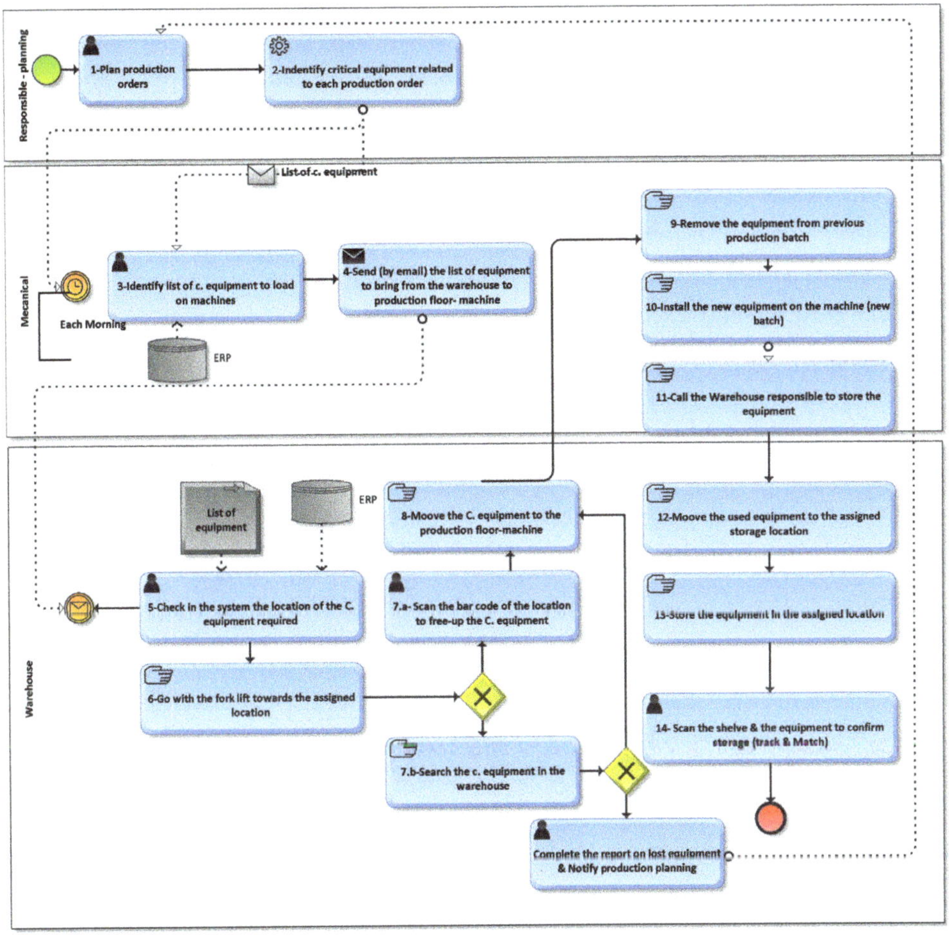

Figure 4. BPMN of the equipment management process.

Therefore, the objective of this pilot project was to develop and test a prototype solution for real-time location of critical equipment (e.g., molds) using the latest passive RFID RTLS available on the market that can (a) be deployed in large multi-shelf warehouses, (b) minimize maintenance management, (c) minimize implementation complexity, and (d) subsequently be used for production management monitoring.

4.2.2. Key Criteria for Our Passive RTLS Prototype

Among the criteria selected for our RTLS application, which justify the relevance of a solution based on passive technologies, are (a) battery life, since molds can be stored for long periods of time before being reused, (b) localization accuracy of the order of one meter, which is sufficient to track a mold on a shelf, (c) the required reading range of 15–20 m from the ground, (d) 3D localization to identify molds on rows of multi-story shelves, (e) the complexity of installation, where we want to minimize cabling and the possibility of using RFID readers powered by POE (Power over Ethernet), with the possibility of relaying the captured information via WIFI, (f) the scalability of the system for use in other areas and/or items to be located, and (g) the total cost of the solution, including acquisition, installation and maintenance costs, installation, and maintenance costs—in this case, passive RFID RTLS solutions stand out for their low-cost tags and zero maintenance. Furthermore, the ability of the new RTLS readers to be installed in high ceilings makes it possible to cover large areas (30–40 m in diameter per reader), thus minimizing the number of readers required.

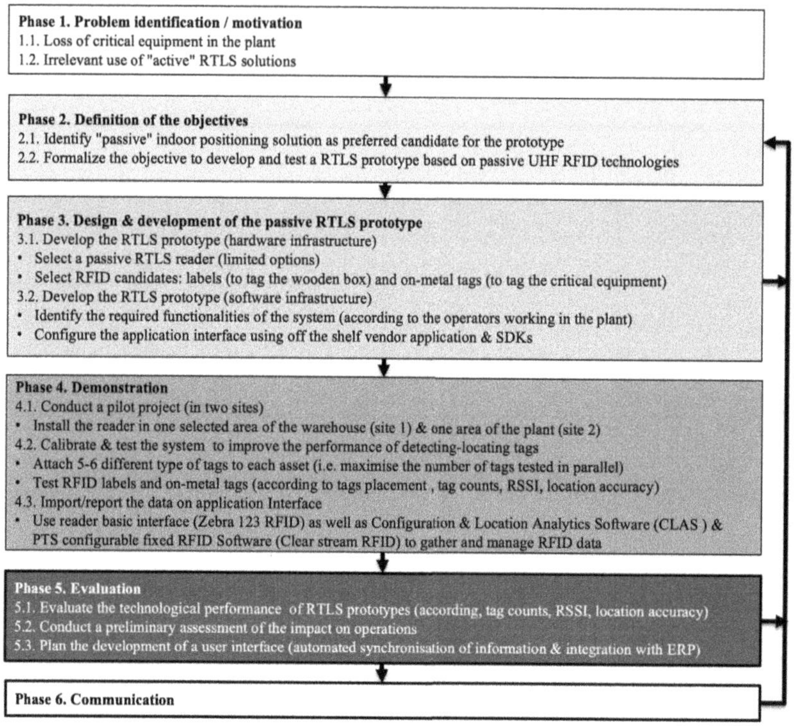

Figure 5. "Design Science" methodology used to build the case.

4.3. Design and Development of Passive RTLS Prototype

During the design and development of the solution (phases 2 and 3), it was first necessary to determine the desired functionalities of the RTLS system: (a) authenticate, (b) create/modify an item, (c) locate an item, (d) update the location of an item, and (e) display the location. This involves calculating the location of an item (x,y,z) and converting this spatial data into useful data for workers (row, shelf).

The IoT infrastructure used to build the RTLS prototype consists primarily of a physical data collection infrastructure based on Zebra's ATR 7000 reader (Zebra Technologies: Lincolnshire, IL, USA) equipped with multibeam RF antennas that cover a 360-degree area. Because of the Zebra infrastructure already installed at the focal company and the familiarity of the employees with Zebra products, the ATR 7000 was chosen over the equally powerful RF Controls reader. Several tags were tested for identifying cases and molds. For cases, we chose Zebra's ZBR4000 model for the crates (where the molds were stored) and Tag Factory's M-Crown (on-metal tag) for the molds (The Tag Factory: Noida, India). Other tags from leading vendors such as Xerafy and Confidex were also tested and performed well. Once data are collected, they are transmitted over the local LAN communications network. The data are hosted in an SQL database to store the last location of each device and the various entities of our solution (SQLight DBMS included in the Django environment for the prototype versus the more robust MySQL or PostgreSQL for deployment). For the software infrastructure/application layer, the use of software development kits (SDKs) provided by the vendor enabled data acquisition and visualization. First, we used the CLAS (Configuration and Location Analytics Software V2.2.45.99) middleware provided by Zebra under license and hosted on a Linux virtual machine (running Ubuntu operating system 20.04) on a local server. This software is used to manage the readers, interpret the data they collect, and visualize the location of the tags on the floor.

4.4. Demonstration of the RTLS Prototype

4.4.1. Testing the Passive RTLS Prototype—Site 1

To test our prototype solution, we conducted a pilot project at the focal company's warehouse. A specific area of the warehouse where the equipment is stored was chosen for the installation of the ATR7000 reader (Figure 6). The reader was mounted on the ceiling to cover the entire area and locate the tagged equipment (1). Selected RFID labels and on-metal tags were attached in various locations on the wooden boxes (2) and the metal molds (3). We were then able to test the (a) reading distance and (b) the reading reliability. For instance, regarding the reading distance, tag RSSI and tag count were used to ensure that the tags were still in the field of view of the reader (4). At this step, the tactic was to gradually reduce the power of the reader and observe which tags continued to respond. At lower power levels, only the "best-performing" tags, with the "best" tag placement were retained. For localization accuracy, the CLAS software was used to view (a) the tag movement on the 2D map and (b) the location accuracy (5).

The pilot project revealed that the range of this passive RTLS reader could reach an actual distance of up to 20 m radius (beyond the manufacturer's specifications)—which means that the range of a reader could cover 40 m of a perimeter in an open space (with a reader 15 m above the ground). In terms of location accuracy, we found that the absolute coordinates returned by the CLAS middleware did not reflect reality. However, the relative position of a tag with respect to the reader and other tags was always correct. The value returned was stable and did not change with each reading, ensuring system reliability. This allowed us to obtain a fairly accurate location of the tools in our RTLS portal, clearly identifying a location column, but without being able to provide the floor (Figure 6).

From a user perspective, we then developed an RTLS application (back end) using the Python programming language to retrieve the raw data from CLAS (x, y, z coordinates) and convert it into more understandable and intuitive information for end-users (i.e., converted into location code). Finally, a web interface (front end) developed using web languages (HTML and CSS) was used to visualize the data provided by the application layer Figure 7.

4.4.2. Testing a Passive RTLS Prototype—Site 2

From a technology performance perspective, although the deployment of our passive RTLS prototype was conclusive, we decided to continue the experimentation in another, more demanding environment. A pilot project was then conducted at the site of a North American manufacturer specializing in the production of fasteners. The manufacturer was struggling with a lack of visibility of work in processes tracking (WIP) of metal buckets used to transport production from zone to zone. Regarding the hardware infrastructure, we also installed an ATR 7000 from Zebra in the staging area where a great number of buckets are waiting to be processed (Figure 8). Regarding the software infrastructure, in this pilot project, all the data were managed using Clearstream RFID middleware v6.0 (Portable Technology Solutions, Calverton, NY, USA) a configurable fixed RFID and Bluetooth beacon software and that can be set up to track asset tags and integrate the data in the ERP of the company. Various information automatically gathered during the tests allows the researchers to assess the performance of each setting (i.e., Tag EPC, tag count, RSSI, date-time, tag-event, etc., Figure 8). To our surprise, in this pilot project, we realized that the tags tested in the previous case were not efficient at all. After various testing (tags, tags placement, reader configuration), we selected Xerafy (Versa Trak, Xerafy: Singapore) and

HID (Exo InLine S, HID: Austin, TX, USA) to maximize the performance of the reads for locating and tracking the buckets.

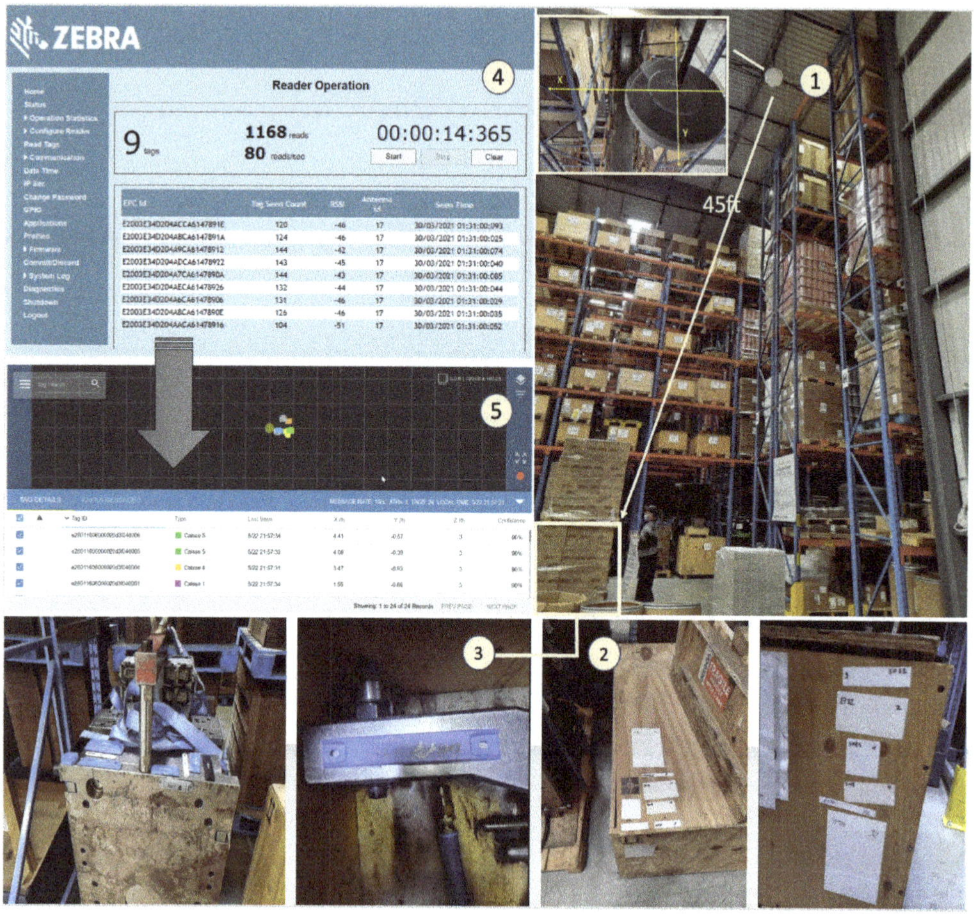

Figure 6. Deployment of ATR 7000 high ceiling passive UHF reader (from Zebra).

The pilot project revealed that the range of this passive RTLS reader could reach an actual distance of up to 10 m radius (or 20 m of a perimeter) in an open space (with a reader 8 m above the ground). This was less than in the previous case, but this is mainly due to the installation height of the reader.

4.5. Evaluation of the RTLS Prototype

In this research project, we have demonstrated that our passive RTLS artifact provides a solution to our initial problem. The impacts are elaborated in the next sections.

4.5.1. Passive RTLS: Evaluation of the Technological Performance

Although in both cases we were able to detect the tagged items, both pilot projects clearly revealed that passive RTLS deployments have to be customized based on each environment. Indeed, the choice of reader location and configuration, as well as the choice of tags, is clearly defined according to the products intended to be located and the operating environment. Both cases, and case 2 in particular, highlighted the interference caused by radio waves (reflection, diffraction) in metallic environments and the difficulty of identifying tags that are not in the reader's line of sight. This effect can be attenuated by combining two readers, as recommended by passive RTLS vendors (vs one reader). This tactic would also have allowed for better coverage and detection of equipment in the blind spots observed during our pilot projects. However, due to the project's budget limitations and the lack of availability of such readers in this market (still new at the time of the pilot), we were unable to obtain a second reader in time.

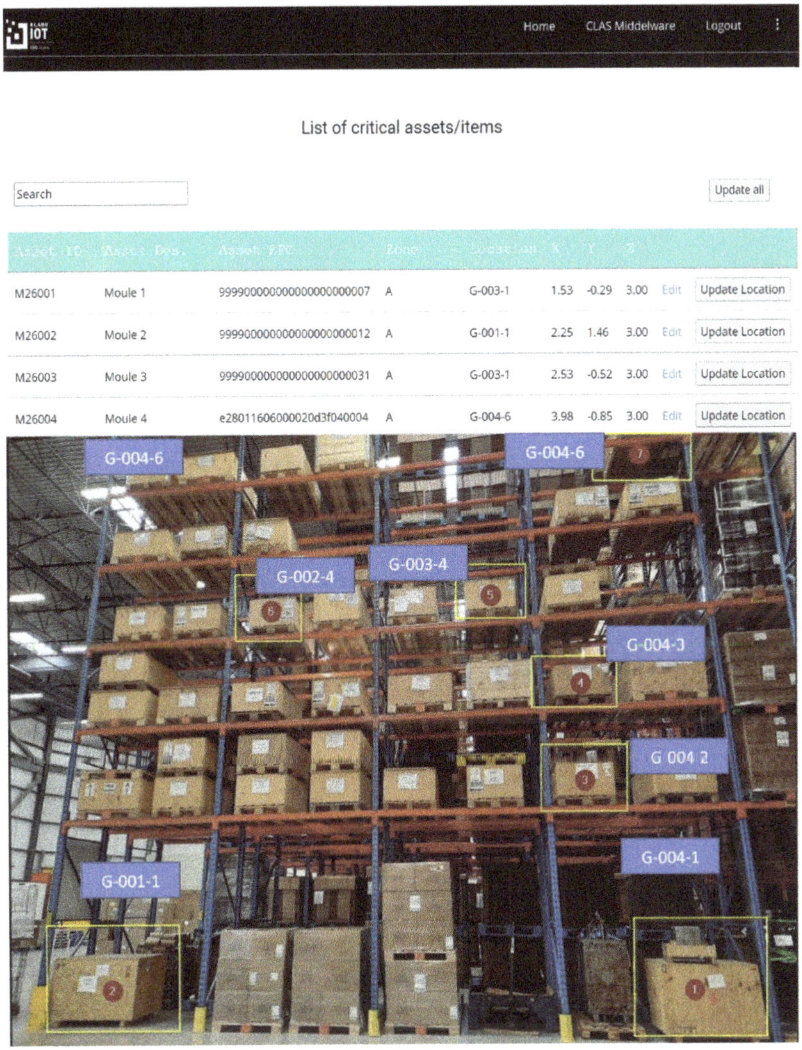

Figure 7. User interface reflecting the "real" location of the equipment.

On the other hand, when it comes to tracking wooden boxes (case 1), the detection was much simpler, especially as special tags have now been developed for use in conjunction with high ceiling readers (i.e., Zebra ZBR4000 tag).

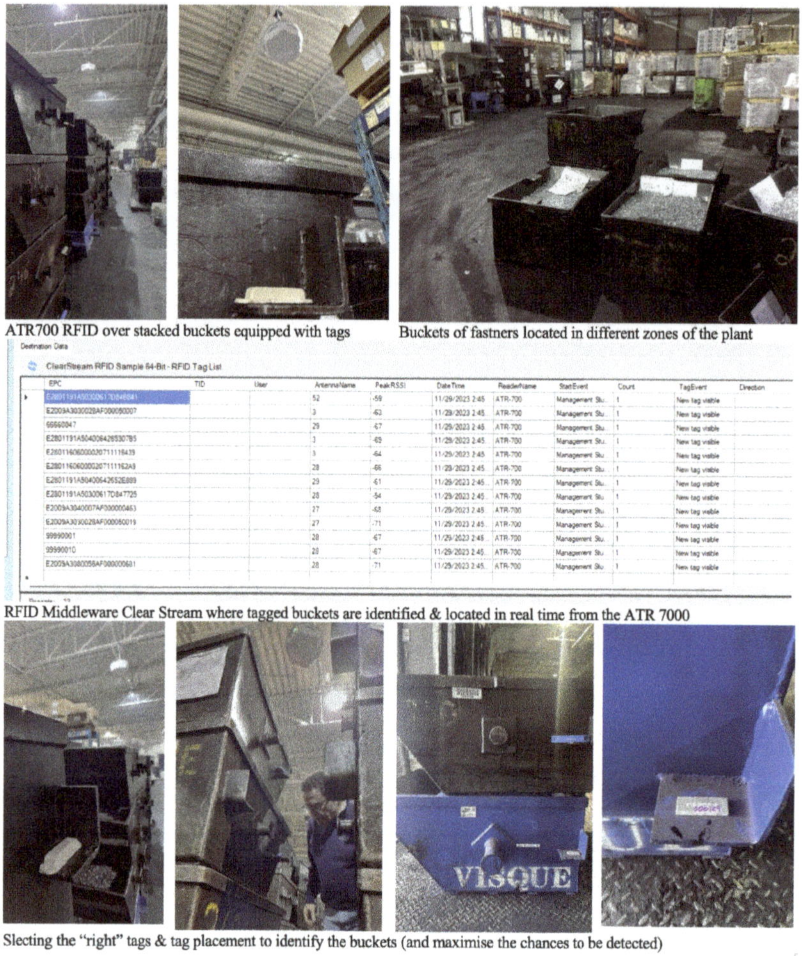

Figure 8. Deployment of ATR 7000 high ceiling passive UHF reader in another manufacturing facility.

4.5.2. Passive RTLS: Evaluation of the Impacts on the Operations

Getting back to case 1, from a process perspective, the solution we implemented automates the manual search for molds, significantly reducing the search time. In fact, an operation that used to take an average of eight hours (up to several days) is now completed in less than 15 min. Given that the cost of a production delay is estimated to be around USD 5000 per hour, this means that each mislocated part can cost up to USD 40,000. So, despite the relative precision of the proposed solution (in terms of shelf number), since the search process is limited to a very specific zone, the search time has been greatly improved. More interestingly, by tagging the storage boxes and the molds, the proposed solution solves put-away problems related to equipment assigned to a "wrong" box. This "untraceable" equipment, unless all the crates are opened, can force the company to re-manufacture a new mold, the machining cost of which is estimated to be around USD 85,000. With tags on both the crates and the molds, the location of the crates and molds (whose tags can be read through the wooden crates) is automatically known. However, to know if molds have been moved to another warehouse, the solution must be deployed in all warehouses, or at least at the warehouse entrances and exits.

From a user perspective, the next step is to have the users (warehouse workers) being able to take ownership of the technology and put it to practical use. This will allow them to evaluate the solution and identify possible gaps/improvements.

From a deployment perspective, the actual prototype has been tested in staging areas and shelving. It would certainly be useful to complement the real-time visibility of these critical assets by tracking them during the production life cycle (e.g., with a track-and-match application). Besides additional passive RTLS readers for wide coverage, fixed RFID readers for checkpoint detection could be used without extensive additional work since they use the same technology, protocols, standards, and platforms.

Finally, from an application point of view, the automated integration of location data with the ERP will make the process of finding equipment more efficient by eliminating the search task in our RTLS prototype. In addition, once the prototype has been validated—as is the case in our two projects—the diversity of IoT platforms compatible with passive RTLS systems greatly simplifies the deployment of such solutions. This should be a definite incentive for companies that decide to go ahead with such innovative projects.

Based on the discussed topics, it is observed that these systems have numerous advantages that could be further explored in future studies. For example, due to the lower cost of tags in comparison to other technologies, they are a suitable option for multi-ownership applications. This means that in a supply chain management system, where a product is transferred from one environment to another with different ownership and management, passive tags can still remain attached to each item. This feature provides current systems with greater flexibility. Therefore, studying how to manage passive systems and integrate them with existing systems is a suitable subject for future research.

5. Conclusions

This joint industry–university project is part of a very strong trend towards digital transformation, where RTLS deployments are still limited. It bridges a gap in the literature by providing a complementary and up-to-date perspective to the scientific literature on RTLS. In this work, we clarified the realistic performance of passive RTLS solutions and proposed a revised typology of RTLS. To achieve this goal, we developed and tested two passive RTLS prototypes using the latest technology available on the market. The pilot projects, conducted (i) in a multi-zone, multi-shelf warehouse location and (ii) in a harsh metal environment with stacked items, allowed us to test the potential and limitations of these latest passive UHF RFID technologies.

With continued innovation and performance improvements, there is no doubt that passive RTLS can be considered a competitor to active RTLS. This can be seen from several perspectives:

- Performance: coverage per antenna is increasing, and performance in terms of reading distance, reliability and location accuracy, low latency is constantly improving.
- Cost: the use of batteryless tags that allow for large-scale deployment with no battery replacement. From a total cost of ownership, (TCO), depending on the area to be covered, multiplying the number of readers can involve a significant upstream investment, but this initial cost can be absorbed by the very low cost of the tags and the maintenance.
- Deployment scalability: the ability of passive RTLS to be deployed in high-ceiling environments makes it suitable for warehouses and factories for both open areas and environments with shelves. In the latter case, however, it is important to ensure that the readers are positioned in the center of the aisles. The minimal configuration and calibration of the readers (vs active technologies), minimal cabling with POE power, and the use of WiFi to facilitate connectivity to servers simplifies the management of such projects.
- Integration: the availability of software SDKs, software applications provided by vendors facilitate the integration with existing systems, and partnerships with channel IoT platform partners are now possible to develop and deploy complete solutions.

Since passive RTLS is now a competitor to active RTLS, managers will have to build their business case accordingly and chose the most appropriate solution. On the other hand, managers can also consider passive and active RTLS as complementary solutions, leveraging the strengths of each system. Indeed, the wide coverage area and the scalability of active RTLS makes them a great option to locate mobile equipment (e.g., forklifts), while passive RTLS can be used to locate and track logistics units in specific zones. To identify the most appropriate RTLS, the decision should be based on the business case requirements including (a) the required functionalities of the RTLS solution, (b) the technical specifications derived from them, and (c) the project constraints.

That said, both pilot projects revealed that passive RTLS solutions are not as straightforward to implement, despite what vendors may suggest. Indeed, each new deployment requires full reevaluation of the RTLS design according to the physical environment.

In addition, to these previous considerations, we need to address another critical aspect of such IoT technology deployments in manufacturing environments and discuss passive RFID-based RTLS Security Concerns, while RFID-based RTLS has emerged as a powerful solution for real-time tracking and monitoring of various assets, objects and individuals, the widespread adoption of such systems also raises significant security concerns that must be carefully addressed to ensure the reliability and secure operation of these systems. The main security concerns associated with RFID-based RTLS are as follows:

- Unauthorized access and data interception: A major concern with RFID-based RTLS is the risk of unauthorized access to the system and interception of data in transit. Attackers may attempt to gain unauthorized access to the system, either physically or by exploiting vulnerabilities in the communication protocols, in order to collect sensitive location data or inject false information. Intercepting data in transit can compromise the confidentiality and integrity of the information.
- Tag cloning and spoofing: RFID tags used in RTLS systems often contain unique identifiers that enable accurate tracking and identification. However, malicious actors may attempt to clone legitimate tags or spoof their identities to gain unauthorized access or manipulate the operation of the system. This can lead to unauthorized tracking, manipulation of location data, or injection of false information, compromising the reliability and trustworthiness of the system.
- Denial of Service (DoS) attacks: The availability and reliability of RFID-based RTLS are highly dependent on the proper functioning of RFID readers, communication networks, and related infrastructure. Attackers can launch denial-of-service (DoS) attacks to disrupt the normal operation of these components, resulting in service interruptions, loss of tracking capabilities, and potential financial and operational consequences. The resilience of RTLS systems to such attacks is critical to maintaining uninterrupted and reliable tracking services.
- Data Integrity and Trustworthiness: Ensuring the integrity and trustworthiness of location data generated by RFID-based RTLS is paramount. Any unauthorized modification or tampering with location data can have serious consequences, such as misinformed decisions or compromised security. Therefore, data authentication, integrity verification, and tamper detection mechanisms are essential to ensure the reliability and trustworthiness of the information provided by the system

To address the above security concerns, several solutions have been proposed and are readily available. These solutions include a variety of approaches such as (i) disabling or "killing" tags to protect sensitive data, (ii) enabling "short-range mode" that allows a tag to respond only when the reader is in close proximity, and (iii) the use of various advanced security protocols (e.g., password) and techniques (e.g., Physically Unclonable Function (PUF)). However, these solutions are not yet readily available or suitable for passive UHF RTLS readers.

Potential technological advancements in Passive RTLS technology may include new tag designs for increased portability and application versatility. Integration of advanced sensors within tags could enable richer data collection, facilitating more precise tracking and environmental monitoring. Enhanced signal processing algorithms could further improve localization accuracy and reliability in complex environments. Additionally, advancements in communication protocols and network architectures could enhance scalability and interoperability, allowing for seamless integration with existing infrastructures. These advancements promise to revolutionize Passive RTLS technology.

Therefore, it is critical to carefully evaluate and weigh the pros and cons of each available solution to determine the most appropriate approach to address the specific security concerns in a given context. By weighing the pros and cons, organizations can make informed decisions to implement the most effective security measures while minimizing potential unintended consequences.

Author Contributions: Y.B. (Yacine Berrabah), S.R., Y.B. (Ygal Bendavid), N.B. and M.S. contributed equally to this work. All authors have read and agreed to the published version of the manuscript.

Funding: This research was funded by Canada Foundation for Innovation (Grant number 40115) and from the Ministère de l'Économie, de l'Innovation et de l'Énergie.

Institutional Review Board Statement: Not applicable.

Informed Consent Statement: Not applicable.

Data Availability Statement: Data are contained within the article.

Conflicts of Interest: The authors declare no conflicts of interest.

References

1. Leng, J.; Sha, W.; Wang, B.; Zheng, P.; Zhuang, C.; Liu, Q.; Wuest, T.; Mourtzis, D.; Wang, L. Industry 5.0: Prospect and retrospect. *J. Manuf. Syst.* **2022**, *65*, 279–295. [CrossRef]
2. Schaeffer, E. *Industry X.0: Realizing Digital Value in Industrial Sectors*; Kogan Page: London, UK, 2017.
3. Affogbolo, R.; Gauzente, C.; Kuntz, P.; Guénoche, A. Bénéfices et risques de l'IdO: Que pouvons-nous en apprendre à propos du changement du Business Model? *Syst. D'inf. Manag.* **2020**, *25*, 59–105. [CrossRef]
4. de Paula Ferreira, W.; Armellini, F.; de Santa-Eulalia, L.A.; Thomasset-Laperrière, V. Extending the lean value stream mapping to the context of Industry 4.0: An agent-based technology approach. *J. Manuf. Syst.* **2022**, *63*, 1–14. [CrossRef]
5. Utterback, J.M. *Mastering the Dynamics of Innovation*; Harvard Business Review Press: Harvard, MA, USA, 1996.
6. Katoch, R. IoT research in supply chain management and logistics: A bibliometric analysis using vosviewer software. *Mater. Today: Proc.* **2022**, *56*, 2505–2515. [CrossRef]
7. Casella, G.; Bigliardi, B.; Bottani, E. The evolution of RFID technology in the logistics field: A review. *Procedia Comput. Sci.* **2022**, *200*, 1582–1592. [CrossRef]
8. Garms, F.; Jansen, C.; Schmitz, C.; Hallerstede, S.; Tschiesner, A. Industry 4.0 Capturing Value at Scale in Discrete Manufacturing. Available online: https://rb.gy/6styig (accessed on 2 January 2024).
9. Kamble, S.S.; Gunasekaran, A.; Gawankar, S.A. Sustainable Industry 4.0 framework: A systematic literature review identifying the current trends and future perspectives. *Process Saf. Environ. Prot.* **2018**, *117*, 408–425. [CrossRef]
10. Industry 5.0—Towards a Sustainable, Human-Centric and Resilient European Industry. Available online: https://research-and-innovation.ec.europa.eu/news/all-research-and-innovation-news/industry-50-towards-more-sustainable-resilient-and-human-centric-industry-2021-01-07_en (accessed on 27 April 2023).
11. Valette, E.; Bril El-Haouzi, H.; Demesure, G. Industry 5.0 and its technologies: A systematic literature review upon the human place into IoT- and CPS-based industrial systems. *Comput. Ind. Eng.* **2023**, *184*, 109426. [CrossRef]
12. Gladysz, B.; anh Tran, T.; Romero, D.; van Erp, T.; Abonyi, J.; Ruppert, T. Current development on the Operator 4.0 and transition towards the Operator 5.0: A systematic literature review in light of Industry 5.0. *J. Manuf. Syst.* **2023**, *70*, 160–185. [CrossRef]
13. Bendavid, Y.; Hachani, M.W.; Rostampour, S. Design et développement d'un prototype de magasin connecté pour les petites entreprise. *Marché Organ.* **2022**, *45*, 49–80. [CrossRef]
14. de Paula Ferreira, W.; Armellini, F.; De Santa-Eulalia, L.A. Simulation in industry 4.0: A state-of-the-art review. *Comput. Ind. Eng.* **2020**, *149*, 106868. [CrossRef]
15. Wang, J.; Lim, M.K.; Wang, C.; Tseng, M.L. The evolution of the Internet of Things (IoT) over the past 20 years. *Comput. Ind. Eng.* **2021**, *155*, 107174. [CrossRef]
16. Behrendt, A.; de Boer, E.; Kasah, T.; Koerber, B.; Mohr, N.; Richter, G. A Manufacturer's Guide to Scaling Industrial IoT. Available online: https://www.mckinsey.com/capabilities/mckinsey-digital/our-insights/a-manufacturers-guide-to-generating-value-at-scale-with-industrial-iot (accessed on 18 December 2023).
17. Jiang, Z.; Guo, Y.; Wang, Z. Digital twin to improve the virtual-real integration of industrial IoT. *J. Ind. Inf. Integr.* **2021**, *22*, 100196. [CrossRef]

18. Liu, M.; Fang, S.; Dong, H.; Xu, C. Review of digital twin about concepts, technologies, and industrial applications. *J. Manuf. Syst.* **2021**, *58*, 346–361. [CrossRef]
19. Ghadge, K.; Achar, T.; Bhatt, A.; Gurumoorthy, B.; Chakrabarti, A. Indoor positioning of metal parts by fingerprinting using passive RFID. *Procedia CIRP* **2020**, *88*, 60–63. [CrossRef]
20. Haibi, A.; Oufaska, K.; El Yassini, K.; Boulmalf, M.; Bouya, M. Systematic mapping study on RFID Technology. *IEEE Access* **2022**, *10*, 6363–6380. [CrossRef]
21. Newman-Casey, P.A.; Musser, J.; Niziol, L.M.; Shedden, K.; Burke, D.; Cohn, A. Designing and validating a low-cost real time locating system to continuously assess patient wait times. *J. Biomed. Inform.* **2020**, *106*, 103428. [CrossRef] [PubMed]
22. Hayward, S.; van Lopik, K.; Hinde, C.; West, A. A Survey of Indoor Location Technologies, Techniques and Applications in Industry. *Internet Things* **2022**, *20*, 100608. [CrossRef]
23. Farahsari, P.S.; Farahzadi, A.; Rezazadeh, J.; Bagheri, A. A Survey on Indoor Positioning Systems for IoT-Based Applications. *IEEE Internet Things J.* **2022**, *9*, 7680–7699. [CrossRef]
24. Zafari, F.; Gkelias, A.; Leung, K.K. A Survey of Indoor Localization Systems and Technologies. *IEEE Commun. Surv. Tutor.* **2019**, *21*, 2568–2599. [CrossRef]
25. Halawa, F.; Dauod, H.; Lee, I.G.; Li, Y.; Yoon, S.W.; Chung, S.H. Introduction of a real time location system to enhance the warehouse safety and operational efficiency. *Int. J. Prod. Econ.* **2020**, *224*, 107541. [CrossRef]
26. Ahmed, F.; Phillips, M.; Phillips, S.; Kim, K.Y. Comparative Study of Seamless Asset Location and Tracking Technologies. *Procedia Manuf.* **2020**, *51*, 1138–1145. [CrossRef]
27. Asaad, S.M.; Maghdid, H.S. A Comprehensive Review of Indoor/Outdoor Localization Solutions in IoT era: Research Challenges and Future Perspectives. *Comput. Netw.* **2022**, *212*, 109041. [CrossRef]
28. Thiede, S.; Sullivan, B.; Damgrave, R.; Lutters, E. Real-time locating systems (RTLS) in future factories: Technology review, morphology and application potentials. *Procedia CIRP* **2021**, *104*, 671–676. [CrossRef]
29. Dai, J.; Wang, M.; Wu, B.; Shen, J.; Wang, X. A Survey of Latest Wi-Fi Assisted Indoor Positioning on Different Principles. *Sensors* **2023**, *23*, 7961. [CrossRef]
30. Ramirez, R.; Huang, C.Y.; Liao, C.A.; Lin, P.T.; Lin, H.W.; Liang, S.H. A practice of BLE RSSI measurement for indoor positioning. *Sensors* **2021**, *21*, 5181. [CrossRef]
31. Che, F.; Ahmed, Q.Z.; Lazaridis, P.I.; Sureephong, P.; Alade, T. Indoor Positioning System (IPS) Using Ultra-Wide Bandwidth (UWB)—For Industrial Internet of Things (IIoT). *Sensors* **2023**, *23*, 5710. [CrossRef]
32. Rahman, A.M.; Li, T.; Wang, Y. Recent advances in indoor localization via visible lights: A survey. *Sensors* **2020**, *20*, 1382. [CrossRef] [PubMed]
33. Zimmerman, T.; Zimmermann, A. Gartner Report, Magic Quadrant for Indoor Location Services. 2023. Available online: https://www.gartner.com/en/documents/4102199 (accessed on 27 April 2023).
34. Zimmerman, T.; Zimmermann, A. Gartner Report, Magic Quadrant for Indoor Location Services. 2022. Available online: https://ubisense.com/gartner-mq-2022/ (accessed on 27 April 2023).
35. Swedberg, C. Impinj Announces Commercial Availability of Its xArray UHF Reader. 2014. Available online: https://www.rfidjournal.com/impinj-announces-commercial-availability-of-its-xarray-uhf-reader (accessed on 12 January 2024).
36. Swedberg, C. RTLS Reader from RF Controls Features Steerable Array Antenna. 2019. Available online: https://www.rfidjournal.com/rtls-reader-from-rf-controls-features-steerable-array-antenna (accessed on 12 January 2024).
37. Swedberg, C. Pilot Aims to Prove Passive RTLS Success for Manufacturing Site. 2019. Available online: https://www.rf-controls.com/wp-content/uploads/2022/09/rfid-journal-article-09062019.pdf (accessed on 12 January 2024).
38. Swedberg, C. Tego Offers IoT Systems with RF Controls' Passive RTLS. 2022. Available online: https://rf-controls.com/wp-content/uploads/2022/10/10102022-Tego-Offers-IOT-Systems-with-RF-Controls.pdf (accessed on 12 January 2024).
39. Swedberg, C. Distance Encoder Automates RFID Tag Writing. 2023. Available online: https://rf-controls.com/distance-encoder-automates-rfid-tag-writing/ (accessed on 14 January 2024).
40. Swedberg, C. New Zebra Reader Promises Real-Time RFID Data Capture. 2019. Available online: https://rainrfid.org/new-zebra-reader-promises-real-time-rfid-data-capture/ (accessed on 14 January 2024).
41. Swedberg, C. Crossdock Solution to Provide Real-Time Location via UHF RFID. 2020. Available online: https://www.rfidjournal.com/crossdock-solution-to-provide-real-time-location-via-uhf-rfid (accessed on 14 January 2024).
42. Mennekens, J. Energy Harvesting Explained. Available online: https://uwinloc.com/wp-content/uploads/2022/04/article-2_jan.pdf (accessed on 17 January 2024).
43. May, M.C.; Glatter, D.; Arnold, D.; Pfeffer, D.; Lanza, G. IIoT System Canvas — From architecture patterns towards an IIoT development framework. *J. Manuf. Syst.* **2024**, *72*, 437–459. [CrossRef]
44. Peffers, K.; Tuunanen, T.; Rothenberger, M.; Chatterjee, S. A Design Science Research Methodology for Information Systems Research. *J. Manage. Inf. Syst.* **2007**, *24*, 45–77. [CrossRef]

sensors

MDPI

Article

Influence of Flexible and Textile Substrates on Frequency-Selective Surfaces (FSS)

Olga Rac-Rumijowska [1,*], Piotr Pokryszka [1], Tomasz Rybicki [2], Patrycja Suchorska-Woźniak [1], Maksymilian Woźniak [1], Katarzyna Kaczkowska [1] and Iwona Karbownik [2]

1 Faculty of Microsystem Electronics and Photonics, Wrocław University of Science and Technology, Wybrzeże Wyspiańskiego 27, 50-370 Wrocław, Poland; piotr.pokryszka@pwr.edu.pl (P.P.); patrycja.suchorska-wozniak@pwr.edu.pl (P.S.-W.)

2 Faculty of Electrical, Electronic, Computer and Control Engineering, Technical University of Łódź, Żeromskiego 116, 90-924 Łódź, Poland; tomasz.rybicki@p.lodz.pl (T.R.); ivakabari@gmail.com (I.K.)

* Correspondence: olga.rac-rumijowska@pwr.edu.pl

Abstract: Frequency-selective surfaces (FSS) are two-dimensional geometric structures made of conductive materials that selectively transmit or reflect electromagnetic waves. In this paper, flexible FSS made on textile and film substrates is presented and compared to show the effect of the texture associated with the type of substrate on the shielding properties. Three geometries of patterns of squares in the border, inversion of squares in the border, and circles with a border were used, and the patterns were made by the silver paste screen printing technique. Microscopic analysis (SEM and optical) was performed to determine the degree of substrate coverage and the actual geometry of the pattern. The resistance per square of the obtained patterns was about 50 mΩ/□. The shielding properties of FSS were simulated in Comsol Multiphysics 6.2 software and then measured by the antenna method. Selective textile filters were obtained, depending on the pattern used, with one or two modals with a transmission attenuation of about 15 dB. The paper analyzes the effect of the substrate and the screen printing technique used on the shielding properties of the flexible FSS.

Keywords: frequency-selective surfaces (FSS); electroconductive textiles; screen printing

1. Introduction

Frequency-selective surfaces (FSS) are periodic arrangements of elementary cells in the form of thin dielectric sheets covered with an electrically conductive pattern (quasi 2D) or three-dimensional elements (3D), which exhibit specific frequency selectivity for electromagnetic waves incident on them [1]. The frequency characteristics of such a structure are completely determined by such parameters as the properties of the dielectric substrate (thickness, dielectric permeability), the properties of the conductive material (layer thickness), the surface geometry, the shape and size of a single element (an elemental cell), and the distance between elements on the surface plane. Classical materials that protect against electromagnetic radiation emitted by electrical equipment are designed to operate over the widest possible range of electromagnetic radiation frequencies [2]. FSS materials, on the other hand, are designed to work only at selected frequencies. However, in both cases, the materials must not let electromagnetic radiation pass through, not only by reflection but also by absorption of the electromagnetic field.

Currently, there is a lot of interest in the creation of new and made-from-new materials FSS structures, their research, and their application as frequency selective filters in the microwave to the optical frequency range. In the literature, there are many geometric formulas for typical frequency-selective material structures and the attenuation/absorption/reflection properties of electromagnetic radiation at the corresponding wavelength assigned to them [3]. For example, shapes with dimensions of 5–50 mm operate in the frequency range of 2.5–15 GHz, and the most common patterns are filled or hollow squares, circles or hexagons, crosses, and square spirals [4]. It is also common to use fractals containing multiple repeated elements of different original sizes to achieve multiband FSS, such as Minkowski and Sierpinski fractals [5].

FSS structures are mostly fabricated on rigid, flat, and smooth surfaces made of laminate [6,7], silicon [8], ceramics [9], or glass [10]. Only in recent years has there been interest in flexible FSS materials [11–13]. Structures of conductive textiles have attracted attention in recent years, partly due to the growing interest in wearable electronics, i.e., clothing that simultaneously acts as sensors, relays, etc., and also due to the advantages such structures offer: lightweight, flexibility, and softness [14,15]. In the literature, one can find isolated examples of research on textile STFs, as their fabrication is difficult and the resulting structures often do not exhibit the desired parameters due to the lack of symmetry, the texture of the substrate, and the difficulty in accurately reproducing the periodic structure [16]. While on rigid substrates this is not a significant technological problem, making structures on fabrics requires excellent process optimization. In printed processes, where the substrate is subjected to significant stresses, its stability, i.e., the non-deformability of the substrate under different conditions of temperature and stress forces, is of primary importance. Such conditions are usually best suited to thin fabrics with low surface weights, high fill (high thread count), and smooth surfaces. These are usually synthetic multifilament silks such as polyester or polyamide. Due to the elasticity and bending of the fabric, it is necessary to use appropriate techniques to cover them with conductive patterns to obtain FSS of good quality. Ibrahim et al. [17] indicate that weaving conductive threads into the fabric structure during the embroidery process makes it possible to obtain FSS of better quality than when covering the fabric with copper foil patterns, whose mechanical properties are completely different from the textile substrate. For this reason, textile STFs are most often made using the conductive fiber/haft weaving technique [14] or screen printing [18]. Even the use of effective and well-studied techniques for applying conductive layers to substrates requires perfect optimization, which, in addition, will be different depending on the parameters of the textile substrate used: the type and thickness of the yarn from which the substrate is made and the weave of the material. However, the shielding results of textile STFs are always worse than the simulation results due to the texture of the substrate. One way to improve the screening performance is to use an intermediate layer that smooths the fabric before the screen printing process [16] or inkjet printing [19]. Applying structures to fabrics by screen printing requires defining their shapes, and due to the characteristics of fabric substrates (lack of stiffness and surface smoothness), the possible FSS structures are designed primarily for microwave radiation [20]. Thus, fabrics made of multifilament synthetic silks (PET) are needed for screen printing. The preparation of the fabrics must include processes of washing (removal of grease and impurities), drying, and thermal stabilization to ensure the required properties of the fabric, including, above all, its resistance to secondary deformation, smoothness, and uniformity of the surface.

The ability to apply different shapes makes it possible to make narrow-band single [21], dual [22], or even quad-mode [23] FSS filters. With a properly designed resonator geometry, it is possible to obtain a flexible FSS, the efficiency of which will depend only slightly on the polarization of the incident electromagnetic radiation and the change in the shape of the printed sample [22]. This indicates the possibility of using textile systems on clothing or making Meta-Skin FSSs [24], which can be used as a masking element for irregular surfaces.

This paper presents a flexible FSS made by the screen printing technique. Identical structures were made on synthetic substrates—polyethylene terephthalate (PET), but to determine the influence of the texture of the substrate on the quality of the print and the mapping of the pattern geometry, two materials were used: Melinex film with a smooth surface and PET polyester fabric. In addition, the results obtained were related to literature data on identical patterns made on rigid substrates. The results of simulations and tests of the shielding effect of the obtained structures in the microwave range are presented.

2. Materials and Methods

2.1. Materials

In this work, the following two types of synthetic substrates of polyethylene terephthalate (PET) were used: textiles and foil. The PET textiles were made by WISTIL S.A., a company from Kalisz, Poland. Pre-washed and heat-stabilized (hot air, 20 s, 190 °C) PET silk-type polyester fabric with a surface weight of 80 g/cm^2 was used for the study. The weight was determined according to PN-ISO 3801:1993—Textile and Fabric Determination of Linear and Surface Mass [25]. Textiles were made in 1/1 plain weave with a thickness of ca. 0.20 mm. The thickness was determined according to EN ISO 5084: 1999—Textile Determination of the Thickness of Textile Products [26]. The average size of macropores in the textiles ranged from 30 to 70 μm.

PET foil, known under the trade name Melinex or Mylar (Selmex company, Plewiska, Poland), is transparent, has a 0.25 mm thickness, and has a surface mass of 349 g/m^2. Both substrates have high resistance to deformations.

A conductive DuPont screen printing paste from the Intexar series was used, intended for flexible substrates, including textile PE874. Resistivity <50 mΩsq for 25 μm thickness; resistivity after crease (ASTM F1683, 180 deg, 1 cycle, 2 kg) <5%; abrasion resistance (ASTM D3363 Pencil Hardness) 1H; adhesion (tape cross-hatch); and the color is metallic.

2.2. Preparation of Frequency-Selective Surfaces

The layers were made by screen printing on a semi-automatic screen printer, Uniprint Go3V (PBT WORKS, Lesni, Czech Republic). A 200-mesh steel screen was used to make the prints, on which a photosensitive emulsion with a thickness of 20 μm was applied. In both substrates, i.e., textiles and foil, the layers were applied using PE874 DuPont paste and by printing from one to four layers. After printing, the layers were dried for 15 min at 130 °C in a well-ventilated dryer (Brucker, Peoria, IL, USA). Tree types of prints were made on each substrate, the shape and dimensions of which are shown in Figure 1a–c. The shape of the patterns was taken from the literature in which Shukor et al. [27] presented simulations and execution of FSS on rigid substrates: FR4 and glass. The entire conductive pattern consisted of periodically arranged unit cells (Figure 1d) and consisted of a matrix: 4 × 4 squares in the border (Figure 1a), 4 × 4 inversion squares in the border (Figure 1b), and 6 × 6 squares in the border.

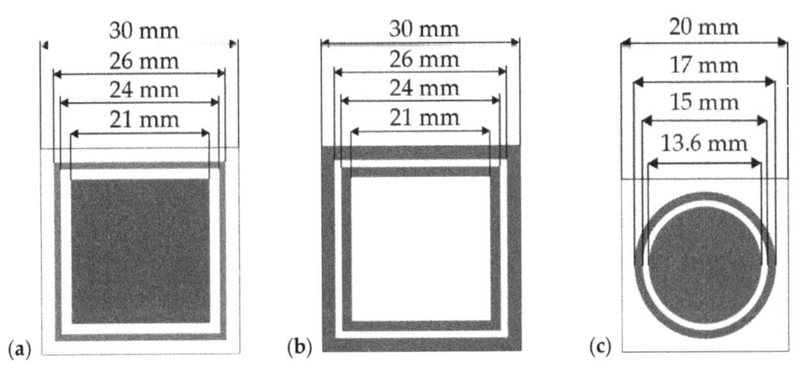

Figure 1. *Cont.*

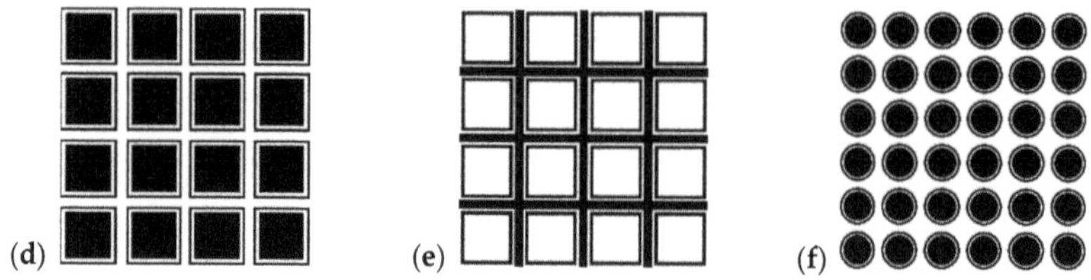

Figure 1. Masks used to make patterns on the screens: (**a**) squares in the border; (**b**) inversion of squares in the border; (**c**) circles with a border; (**d–f**) FSS made from elements (**a–c**).

2.3. Characterization

The morphology of FSS was determined by the scanning electron microscope Hitachi SU6600 (Tokyo, Japan) and the optical microscope Leica DM4500 B LED (Wetzlar, Germany). To determine the geometry of the printed elements on the textile substrate and film, microscopic images were taken of all the elements at different locations on the sample, and each value was measured a minimum of 10 times using the microscope software LAS X for Leica DM4500 B LED. The study of electrical conductance was a 2-point method. Gold-plated, spring-loaded measurement probes with a diameter of 35 mm and rounded ends were selected for the measurements, which enabled non-destructive measurements and ensured good electrical contact. The probes were placed on a rigid plate, and the whole thing was loaded with two 500 g weights. Before the measurements, the textiles were stored in an atmosphere with a relative humidity of 60% at a temperature of 20 °C for 24 h. The tests were carried out at a constant temperature of 20 °C in an ambient atmosphere with a relative humidity of 30%. DC tests were performed using the Keithley Instruments 610C Solid-State Electrometer device (Cleveland, OH, USA).

The resistance of the printouts was measured ten times and averaged. On this basis, the resistance per square R□ was calculated according to Formula (1).

$$R\square = R_{av}/n \quad (1)$$

where R□—surface resistance (resistance per square), R_{av}—average resistance, and n—number of squares (assumed surface units).

Electromagnetic simulation of FSS structures was performed using the finite element method (FEM) with Comsol Multiphysics software version 5.4 and the RF (Radio Frequency) module. The simulations analyzed the parameters of the scattering matrix S, such as S21, which contains information about the transmission of the electromagnetic wave propagated from the source to the receiver, as well as S11, which carries information about the reflected signal from the tested structure to the source. Simulations were carried out for a substrate made of 0.2 mm thick film and 0.2 mm thick fabric with dielectric permeabilities (εr) of 3.4 [28] and 1.9 [29], respectively. Silver with a resistivity of 50 mΩsq was used as the metallic layer. Geometrical dimensions are given in Figure 1. Electromagnetic simulations were performed for a single elementary cell in the frequency range from 1 to 18 GHz. During the analysis of S-parameters, a virtual matrix of FSS structures of 10 × 10 elements was used to analyze the interaction of an electromagnetic wave with a periodic structure.

Shielding properties of FSS were carried out at the Accredited Laboratory for Electromagnetic Compatibility and Electromagnetic Field Measurements (LBEMC), where transmission (S21) measurements were made in the frequency range from 1 to 18 GHz using the antenna method. For this purpose, we used a N5245A vector analyzer from Agilent Technology (Santa Clara, CA, USA) operating in the frequency range of 10 MHz to 50 GHz, an ETS-Lindgren 3115 Horn Antenna (Lindgren, Eura, Finland), enabling measurements from 1 to 18 GHz, and a carbon absorber with a high electromagnetic absorption coefficient of 0.5–40 GHz cat no. MWA-LFA-610 × 610 × 50.8-O (MTC Micro Tech Components, Dillingen an der Donau, Germany). Figure 2 shows (a) a photo and (b) a schematic of the test stand used in the measurement of S-scattering parameters.

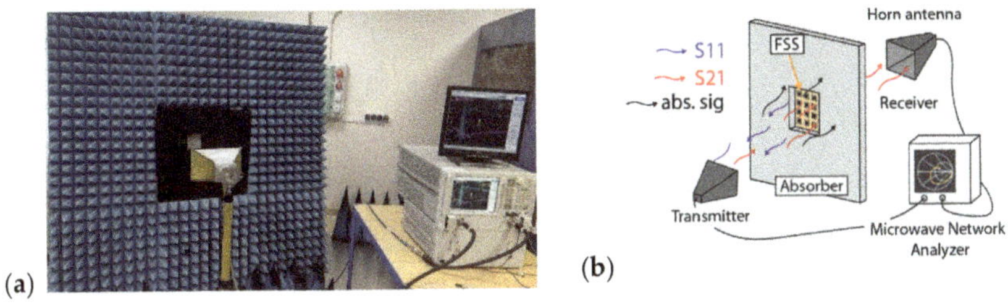

Figure 2. Measurement stand for testing electromagnetic radiation transmission by antenna method: (**a**) photo of the measurement stand; (**b**) schematic diagram with description of the measurement stand.

3. Results and Discussion

3.1. Morphology of Frequency-Selective Surfaces

The performance of the fabricated structures having the character of selective band-stop filters largely depends on the quality of the geometric patterns made and the resistivity of the printing, which is closely related to good coverage of the fabric with conductive paste. Images from a scanning electron microscope (SEM) showed that only the printing of three conductive layers ensures complete coverage of the fabric structure (Figure 3). With fewer layers, the pattern produced is discontinuous, and uncovered fibers are still visible on the surface. However, even when the fabric is fully covered, its characteristic weave and surface irregularity are visible, and there is more paste in the recesses than on the convex parts. Moreover, even the continuous layer has a distinctly grainy and heterogeneous structure (Figure 3c, inserted image).

Figure 3. SEM images of PET textile covered by: (**a**) one, (**b**) two, and (**c**) three layers of PE874 paste (inserted image at a higher magnification).

The screen printing process is characterized by relatively high printing inaccuracy, which further depends on the type of substrate used. In the case of a textile substrate (Figure 4c), a stepped edge of the print is visible, caused by the placement of the yarn in two directions perpendicular to each other. On the other hand, in the case of the substrate made of foil (Figure 4d), the paste melting at the edge of the print is visible. The blurring seen on the film is a result of the low viscosity of the screen printing paste, which is designed to penetrate well into the textile fabric. For this reason, in many cases, structures made on films had larger deviations from the designed dimensions than structures made on textile substrates, which was confirmed by measurements of all structures (both printed elements and the spaces between them) taken on microscopic images and included in Table 1. The difference in the widths of the individual elements and the gaps between them were in the range of 0.5–17% and 5–23%, respectively, with the gaps between the prints always smaller than designed and the prints smaller or larger (Table 1). These differences are typical values for standard thick-layer technology.

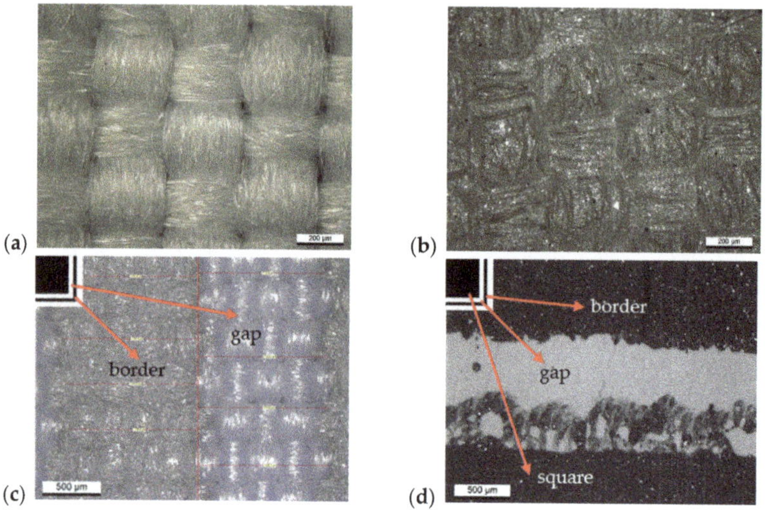

Figure 4. PET textile: (**a**) uncovered and (**b**) covered by three layers. Edges of the squares in the border printed on: (**c**) textiles and (**d**) foils.

Table 1. Comparison of design and measured size of printed elements (T—textiles; F—foil).

Sample	Element	Substrate	Design Size [mm]	Measured Size [mm]
squares in the border	square	T	21	20.0 (+/−0.04)
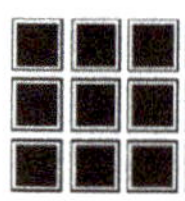		F		19.4 (+/−0.18)
	gap	T	1.5	1.16 (+/−0.04)
		F		1.33 (+/−0.06)
	border	T	1	1.17 (+/−0.04)
		F		0.94 (+/−0.44)
	distance	T	4	3.40 (+/−0.04)
		F		3.56 (+/−0.03)
inversion of squares in the border	border	T	1.5	1.63 (+/−0.04)
		F		1.70 (+/−0.12)
	gap	T	1	0.92 (+/−0.01)
		F		0.89 (+/−0.34)
	line	T	4	4.2 (+/−0.01)
		F		3.98 (+/−0.05)
circles with a border	circles	T	1.36	1.30 (+/−0.06)
		F		1.25 (+/−0.06)
	gap	T	7	5.90 (+/−0.06)
		F		6.59 (+/−0.12)
	border	T	10	9.63 (+/−0.03)
		F		9.89 (+/−0.07)
	distance	T	3	2.71 (+/−0.03)
		F		2.73 (+/−0.04)

3.2. Electrical Resistance

The shielding properties of frequency-selective surfaces depend largely on the resistance of the conductive material used and the quality of the pattern. For this reason, research was carried out to determine the optimal number of layers applied to the textile and foil substrates. The obtained results (Table 2) showed that in the case of fabric, with each layer applied, an increasingly lower resistance per square is obtained, but after applying three layers, it is comparable to the catalog value of 0.050 Ω/□ because then complete coverage occurs of material with conductive paste, as shown by microscopic images (Figures 3 and 4). When using more layers, the resistance per square was lower, but the print quality deteriorated due to the blurring of the pattern. The behavior of the print on Melinex foil is similar. After applying more than three layers, the print was significantly blurred (which, in this case, did not penetrate the fabric), which resulted in a significant increase in resistance. Based on the obtained results, samples of both materials with three layers of silver paste printed on them were selected for testing the shielding properties.

Table 2. Comparison of resistance per square R□ for different numbers of printed layers.

Number of Layers/ Substrate	1 [Ω/□]	2 [Ω/□]	3 [Ω/□]	4 [Ω/□]	5 [Ω/□]
Textile	0.096	0.065	0.057	0.037	0.033
Foil Melinex	0.053	0.050	0.048	0.080	0.090

3.3. Simulation Results

The simulations performed show that the use of a two-element pattern consisting of a filled geometric figure (circle or square) in a border of the same shape makes it possible to obtain a dual-mode selective filter, in which the geometric element with a filled conductive layer is responsible for attenuation of radiation at lower frequencies and the ring is responsible for attenuation in the higher frequency range, which is perfectly evident in the case of simulations of the pattern circles with a border (Figure 5c). On the other hand, making patterns that are reciprocal inversions, as in the case of squares in the border (Figure 5a) and inversions of squares in the border (Figure 5b), makes it possible to obtain materials with mutually opposite reflection (S11) and transmission (S21) properties of electromagnetic radiation. However, when designing a multimode filter, it should be borne in mind that a pattern consisting of complex elements, as in the present case, will not behave like the sum of the attenuation of several elements. A structure consisting of squares in the border will not act like the sum of the actions of a structure containing only squares or only borders because both structures interact with each other, and thus the type of interaction depends on the distance between elements. In addition, the distance between elementary cells is also extremely important. When it is too small, it causes the resonance of neighboring rings between each other rather than between a square and a ring. In the case of squares in the border and inversion of squares in the border, the resonance of the interaction of neighboring elements at about 8.5 GHz is visible (Figure 5a), which is not the case in the circles with a border structure. The pattern of circles in a circle is an example of a properly designed structure in which the formation of two narrow bands of attenuation of electromagnetic radiation can be observed.

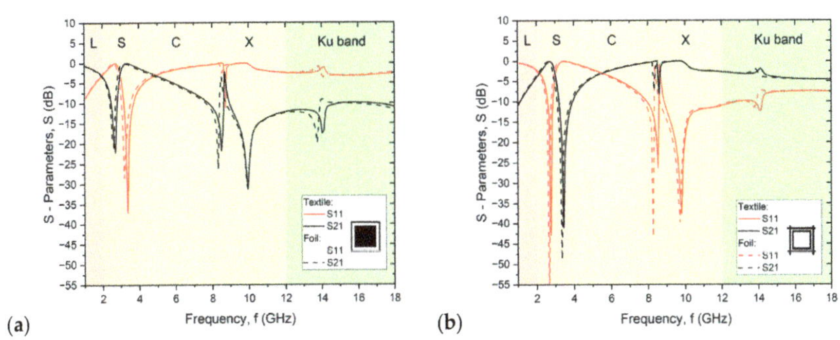

Figure 5. *Cont.*

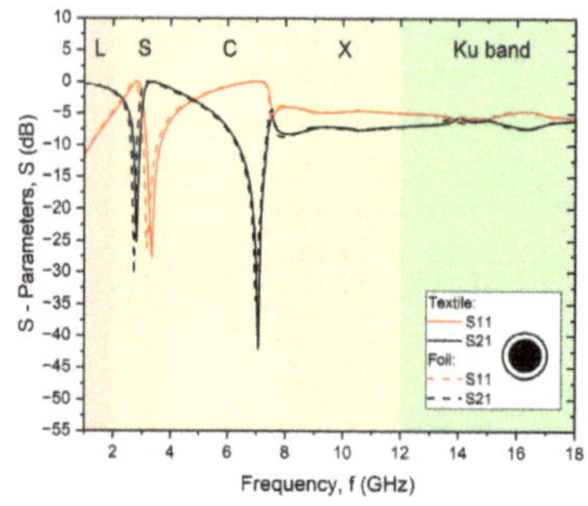

(c)

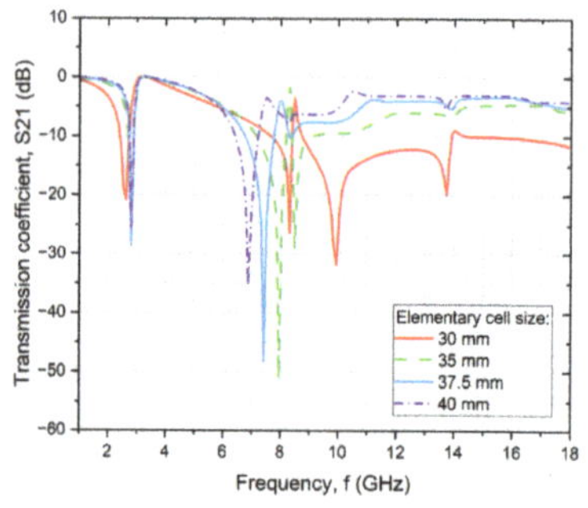

(d)

Figure 5. Simulation results for structures printed on foil and textile: (**a**) squares in the border; (**b**) inversion of squares in the border; (**c**) circles with a border (**d**) changes in elementary cell size for structures (**a**,**b**).

Therefore, a change in the geometric parameters of structures a and b should be taken into account when designing subsequent designs. Concerning the simulation results (Figure 5d), the distances between elementary cells for structures a and b should be increased to a minimum of 37.5 mm to avoid the negative phenomenon of resonance formation instead of a second attenuation peak. In this publication, as mentioned above, the geometry of resonator elements operating at 3.4 GHz fabricated on rigid substrates was adopted to compare their performance [27].

3.4. Shielding Properties

Measurements of the scattering matrix S using the antenna method made it possible to determine the transmission of the tested structure using parameter S21. It was shown that for all patterns (Figure 6), very similar results were obtained regardless of the substrate used. However, in all cases, there is a slight shift in the characteristics obtained for the print on fabric relative to the print on film, which is mainly due to the difference in the dielectric permeability of the substrate affecting the resonant frequency. The thickness of the substrate in both cases was the same, but the substrates differed in surface texture. The high convergence between the results of the two substrates and the discrepancy between the results obtained and the simulations mean that it is the technique used to make the conductive layer and the material from which it was made that determines the effectiveness of the textile FSS.

The lower radiation attenuation values relative to the simulations are also due to the losses associated with Joule heating (in the conductive layers) and dielectric heating (in the substrate) not included in the simulations [30].

For both squares in the border and their inversion in the lower frequency range, a narrow radiation attenuation peak is evident at 2.4 GHz and 3.2 GHz, respectively, whose presence is associated with a full or hollow square, respectively. These peaks coincide quite well with the values obtained from simulations. Their slight offset from the simulated values is mainly due to the difference in the geometry of the patterns given in simulation and those obtained in reality (Table 1, Figure 4). On the other hand, about half of the weaker attenuation of electromagnetic radiation is related to the quality of prints obtained in the screen-printing process and the discontinuity resulting from the composite composition of the paste, compared to the ideal conductor, i.e., a continuous layer of silver.

The size of the circle width, the border, and the distances between elements were significantly smaller than their counterparts in the squares in the border pattern, which made their implementation much more difficult, and the resulting layer was characterized by quite large imperfections affecting the electrical parameters. Changing the real and imaginary parts of the conductor impedance directly affects the goodness of the filter (Q), that is, the bandwidth fBW, the signal transmission level, and the characteristic frequency (f0). Its change causes a shift in the transmission peak, which is most significant in the case of the pattern of circles with a border (Figure 5c). Disturbances in the mapping of the geometry of the structures have a very significant impact on the shielding efficiency and frequency characteristics of the FSS, which is also confirmed by the study of Turki et al. [31], in which FSS structures were made with intentional pattern disturbances in the form of making incomplete squares, of which the FSS structure consisted.

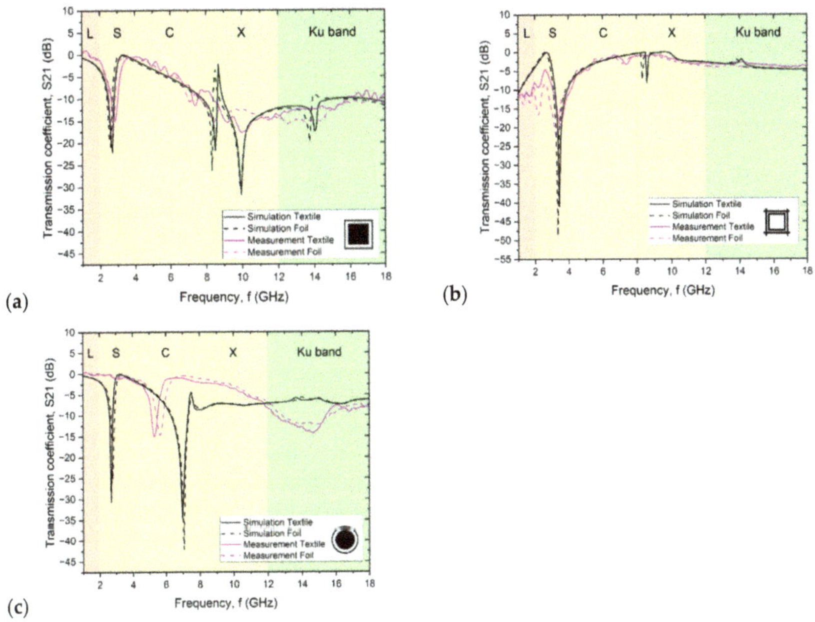

Figure 6. Test and simulation results for structures printed on foil and textile: (**a**) squares in the border; (**b**) inversion of squares in the border; and (**c**) circles with a border.

In the higher-frequency region around 8 GHz in all three structures (Figure 6a–c), the simulations show a second peak associated with the presence of a thin envelope or its inverse, which is absent or much less intense and shifted in the measurement results. These larger deviations of the measured results from the simulated values are the result of a change in the way electromagnetic radiation interacts with the material in the higher frequency range. For the simulations, the parameters of a perfectly smooth layer with equal thickness, constant resistivity, and an electron model of conductivity were assumed. In contrast, screen-printed conductive elements are composite layers consisting of approximately micrometer-sized grains surrounded by polymer. In conductive composite materials, even when the percolation threshold is exceeded, there is always the phenomenon of current tunneling through the thin layer of polymer surrounding the conductive particles. For this reason, even after crossing the percolation threshold, the conductivity of the composite is always 2–4 orders of magnitude lower than that of the filler used [32,33]. As shown by microscopic images (Figure 3), the structure of the layer shows approximately micrometer-sized silver grains, which form a non-uniform, rough layer. In addition, the layer is characterized by high surface heterogeneity due to the shape of the fabric, which is still visible even after three layers were printed (Figure 4b). The inhomogeneity of the layer and the impossibility of 100% reproduction of the assumed shape affect the resistivity of the layer, especially the small and narrow elements such as the borders of squares or circles. There are larger errors in these elements, and they are responsible for the interaction at higher frequencies.

In addition, there is a skin effect in the high-frequency region, causing the current density near the conductor's surface to be greater than inside the conductor. As a result, there is an increase in the impedance of the conductor, causing an increase in the resulting power losses in the conductor. One of the quantities characterizing the skin effect is the depth of current penetration into the conductor, and in the studied range, it is of the order of several micrometers. In the case of the structures studied, the depth of penetration of the electromagnetic wave is comparable to the roughness and unevenness of the surface due to both the properties of the substrate and the specifics of the screen printing technique. As a result, the electromagnetic wave in such a material is scattered, and there is a change in the conductivity of the material, whose impedance, capacitance, and inductive reactance change value. The result is the practical disappearance of the narrow attenuation peak in the high-frequency range and its replacement by a much wider and less intense peak in the Ku band range.

For peaks near 3 GHz, there is no scattering effect of the high-frequency signal, and in addition, the interacting structures in this range are filled circles or squares, which have better filling and surface coverage and thus lower layer resistance. This makes it possible to produce a narrow-band filter in this frequency range of electromagnetic radiation.

4. Conclusions

Textile frequency-selective surfaces utilize textiles as substrates for FSS, offering flexibility, lightweightness, and conformability. This allows them to be used in stealth technology—integrated into clothing or camouflage nets to reduce radar cross-section, making military personnel or equipment less detectable to radar systems; electromagnetic shielding—can be used in clothing or tents to shield against electromagnetic interference (EMI) or electromagnetic radiation (EMR) from sources like cell phones, Wi-Fi routers, or power lines, protecting individuals or sensitive equipment; communication systems—can be employed in wearable antennas for wireless communication devices, such as smartwatches, fitness trackers, or medical monitoring devices, improving signal reception and transmission efficiency; or architectural applications—can be incorporated into building materials, curtains, or window films to control the transmission of specific frequencies of electromagnetic waves, such as blocking harmful UV radiation while allowing visible light to pass through.

Structures containing a square or circular shape in the border are often used in the design of multimode FSSs [3,34,35]. The purpose of this paper was to compare structures fabricated on a textile substrate to analogous structures designed and fabricated on rigid FR4 substrates reported in the literature [27]. In this paper, the reflection (Sl1) and transmission (S21) of the design FSS structure are analyzed. Analyzing the structures with the same design (configuration 1) for the circular loop pattern, two peaks are visible on the transmission spectrum: the first around 1.7 GHz (−35 dB) and the second around 4 GHz, while in the square loop measurements, the first narrow peak is around 1.9 GHz (−26 dB) and the second around 5.5 GHz. In addition, both charts show a general reduction in transmission in the broad spectral range, which occurs already above 3 GHz. This behavior is consistent with our results, in which there are also two peaks on the transmission spectrum and a broad-spectral attenuation of transmission in the high-frequency range is evident. The obtained attenuation for individual peaks in all fabricated structures, both on the textile substrate and on the film, is about 15 dB, or about half that of analogous structures fabricated on the glass-epoxy (FR4) laminate presented in [27]. The simulation results conducted by the authors were consistent with the results available in the literature [27]. However, the actual results obtained from measurements of both foils and fabrics were half as efficient. This is related to the quality of the layers made using screen printing.

The layers made by the screen printing technique are characterized by less continuity than the metallic conductive layers on the laminate, and it is also not possible to reproduce the shape as accurately as in the case of etching the conductive layer in the laminate. While in screen printing it is possible to make structures where the geometric distribution of the elements to each other will be relatively constant, it is more difficult to make elements with assumed geometries and without structure errors within the surface of a single print. Any perturbation of the structure results in the broadening of the peaks, shifting of the peaks, or weakening of the attenuation. In this work, the effect of individual perturbations in the structure of flexible and textile FSSs on their shielding properties is demonstrated. The effects of individual print parameters on the shift of attenuation bands, the reduction or disappearance of electromagnetic radiation attenuation, or the formation of resonances, respectively, are presented.

The structures used in the article and, in particular, squares in the border and the inversion of squares in the border have too small distances between elementary cells. As a result, the resonance between rings in separate elementary cells in the frequency range of 8–10 GHz is visible (Figure 6a,b). Increasing the distance between elementary cells from a value of 30 mm to 37.5 mm for the squares in the border and similarly for the inverted squares should result in the disappearance of the resonance and thus the formation of a two-compartment radiation filter (Figure 5d), which will be the subject of further work by the authors.

Author Contributions: Conceptualization, O.R.-R. and P.P.; investigation, O.R.-R., P.P., M.W., K.K. and P.S.-W.; methodology, O.R.-R. and P.P.; validation, O.R.-R. and P.P.; software, P.P. and T.R.; supervision O.R.-R. and I.K.; writing—original draft preparation O.R.-R.; writing—review and editing P.P., T.R. and I.K. All authors have read and agreed to the published version of the manuscript.

Funding: This work has been supported by project Miniatura no. DEC-2023/07/X/ST7/00027, financed by the Polish National Science Centre (NCN), and funding from the statutory activities of the Wrocław University of Science and Technology (WRUST) Faculty of Electronics, Photonics, and Microsystems. Microwave equipment was purchased from grant POIG.01.01.02-00-008/08-05 (co-financed by the European Union through the European Regional Development Fund), Leader Grant No. 027/533/L-5/13/NCBR/2014 (the National Centre for Research and Development), and TECHMATSTRATEG project No. 1/346922/4/NCBR/2017.

Institutional Review Board Statement: Not applicable.

Informed Consent Statement: Not applicable.

Data Availability Statement: Data are contained within the article.

Acknowledgments: Special acknowledgments to Ryszard Korbutowicz for taking SEM images.

Conflicts of Interest: The authors declare no conflicts of interest.

References

1. Ferreira, D.; Caldeirinha, R.F.; Cuinas, I.; Fernandes, T.R. A Review of Manufacturing Materials and Production Methods for Frequency-Selective Structures. *IEEE Antennas Propag. Mag.* **2018**, *60*, 110–119. [CrossRef]
2. Rybicki, T.; Stempien, Z.; Karbownik, I. EMI Shielding and absorption of electroconductive textiles with PANI and PPy conductive polymers and numerical model approach. *Energies* **2021**, *14*, 7746. [CrossRef]
3. Anwar, R.S.; Mao, L.; Ning, H. Frequency selective surfaces: A review. *Appl. Sci.* **2018**, *8*, 1689. [CrossRef]
4. Kapoor, A.; Mishra, R.; Kumar, P. Frequency selective surfaces as spatial filters: Fundamentals, analysis and applications. *Alex. Eng. J.* **2022**, *61*, 4263–4293. [CrossRef]
5. Gianvittorio, J.P.; Romeu, J.; Blanch, S.; Rahmat-Samii, Y. Self-similar prefractal frequency selective surfaces for multiband and dual-polarized applications. *IEEE Trans. Antennas Propag.* **2003**, *51*, 3088–3096. [CrossRef]
6. Yan, M.; Qu, S.; Wang, J.; Zhang, J.; Zhou, H.; Chen, H.; Zheng, L. A miniaturized dual-band FSS with stable resonance frequencies of 2.4 GHz/5 GHz for WLAN applications. *IEEE Antennas Wirel. Propag. Lett.* **2014**, *13*, 895–898. [CrossRef]
7. Miao, X.G.; Feng, Q.W.; Fang, X.W.; Ghezzo, F.; Zhao, Z.Y.; Liu, R.P. Design, Fabrication and Characterization of Fused Silica-Based Composites with an LTCC-Derived FSS. *Adv. Mater. Res.* **2014**, *941*, 305–309. [CrossRef]
8. Zendejas, J.M.; Gianvittorio, J.P.; Rahmat-Samii, Y.; Judy, J.W. Magnetic MEMS reconfigurable frequency-selective surfaces. *J. Microelectromechanical Syst.* **2006**, *15*, 613–623. [CrossRef]
9. Tang, L.; Wang, Y.; Huang, J. A Thermostable Frequency Selective Surface with both a Low-pass and a Wide Shielding Band. In Proceedings of the 2023 International Telecommunications Conference (ITC-Egypt), Alexandria, Egypt, 18–20 July 2023; IEEE: Piscataway, NJ, USA, 2023; pp. 34–37.
10. Liu, X.; Tan, W.; Shen, Z.; Jin, C. Integrated frequency selective surface and antenna printed on a transparent substrate. *IEEE Antennas Wirel. Propag. Lett.* **2020**, *19*, 2062–2066. [CrossRef]

11. Wang, L.B.; See, K.Y.; Zhang, J.W.; Salam, B.; Lu, A.C.W. Ultrathin and flexible screen-printed metasurfaces for EMI shielding applications. *IEEE Trans. Electromagn. Compat.* **2011**, *53*, 700–705. [CrossRef]
12. Karbownik, I.; Malinowska, G.; Rybicki, E. Textile Multi-layer Systems for Protection against Electromagnetic Radiation. *Fibres Text. East. Eur.* **2009**, *17*, 73.
13. Brzeziński, S.; Rybicki, T.; Malinowska, G.; Karbownik, I.; Rybicki, E.; Szugajew, L. Effectiveness of shielding electromagnetic radiation, and assumptions for designing the multi-layer structures of textile shielding materials. *Fibres Text. East. Eur.* **2009**, *1*, 60–65.
14. Tennant, A.; Hurley, W.; Dias, T. Experimental knitted, textile frequency selective surfaces. *Electron. Lett.* **2012**, *48*, 1386–1388. [CrossRef]
15. Chauraya, A.; Seager, R.; Whittow, W.; Zhang, S.; Vardaxoglou, Y. Embroidered Frequency Selective Surfaces on textiles for wearable applications. In Proceedings of the 2013 Loughborough Antennas & Propagation Conference (LAPC), Loughborough, UK, 11–12 November 2013; pp. 388–391.
16. Ghebrebrhan, M.; Aranda, F.; Walsh, G.; Ziegler, D.; Giardini, S.; Carlson, J.; Gatesman, A. Textile frequency selective surface. *IEEE Microw. Wirel. Compon. Lett.* **2017**, *27*, 989–991. [CrossRef]
17. İbrahim, Ü.N.E.R.; Sultan, C.A.N.; Gürcüm, B.H.; YILMAZ, A.E.; Aksoy, E. Design and Implementation of a Textile-Based Embroidered Frequency Selective Surface. *Text. Appar.* **2022**, *32*, 297–303.
18. Seager, R.D.; Chauraya, A.; Bowman, J.; Broughton, M.; Nimkulrat, N. Fabrication of fabric based frequency selective surfaces (FSS). In Proceedings of the The 8th European Conference on Antennas and Propagation (EuCAP 2014), The Hague, The Netherlands, 6–11 April 2014; IEEE: Piscataway, NJ, USA, 2014; pp. 1978–1980.
19. Whittow, W.G.; Li, Y.; Torah, R.; Yang, K.; Beeby, S.; Tudor, J. Printed frequency selective surfaces on textiles. *Electron. Lett.* **2014**, *50*, 916–917. [CrossRef]
20. Can, S.; Karakaya, E.U.; Yılmaz, A.E. Design, fabrication, and measurement of textile-based frequency selective surfaces. *Microw. Opt. Technol. Lett.* **2020**, *62*, 3444–3450. [CrossRef]
21. Almirall, O.; Fernández-García, R.; Gil, I. Wearable metamaterial for electromagnetic radiation shielding. *J. Text. Inst.* **2022**, *113*, 1586–1594. [CrossRef]
22. Tak, J.; Choi, J. A wearable metamaterial microwave absorber. *IEEE Antennas Wirel. Propag. Lett.* **2016**, *16*, 784–787. [CrossRef]
23. Cınar, A.; Basaran, S.C. Textile-based quad-band electromagnetic absorber design for wearable applications. *J. Text. Inst.* **2023**, 1–5. [CrossRef]
24. Yang, S.; Liu, P.; Yang, M.; Wang, Q.; Song, J.; Dong, L. From flexible and stretchable meta-atom to metamaterial: A wearable microwave meta-skin with tunable frequency selective and cloaking effects. *Sci. Rep.* **2016**, *6*, 21921. [CrossRef] [PubMed]
25. *PN-ISO 3801:1993*; Textile and Fabric Determination of Linear and Surface Mass. Polish Committee for Standardization: Warsaw, Poland, 1993.
26. *EN ISO 5084:1999*; Textile Determination of the Thickness of Textile Products. International Organization for Standardization: Geneva, Switzerland, 1999.
27. Shukor, M.M.; Aziz, M.A.; Ahmad, B.H.; Suaidi, M.K.; Johar, M.F.; Othman, M.A.; Malek, M.A. Characteristic impedance modeling of circular loop and square loop frequency selective surface (FSS) on hybrid material. In Proceedings of the 2014 International Symposium on Technology Management and Emerging Technologies, Bandung, Indonesia, 27–29 May 2014; IEEE: Piscataway, NJ, USA, 2014; pp. 486–491.
28. Available online: https://www.ukinsulations.co.uk/pdfs/Melinex_O.pdf (accessed on 2 February 2023).
29. Dhupkariya, S.; Singh, V.K.; Shukla, A. A review of textile materials for the wearable antenna. *J. Microw. Eng. Technol.* **2015**, *1*, 1–8.
30. Ghebrebrhan, M.; Bermel, P.; Yeng, Y.X.; Celanovic, I.; Soljačić, M.; Joannopoulos, J.D. Tailoring thermal emission via Q matching of photonic crystal resonances. *Phys. Rev. A* **2011**, *83*, 033810. [CrossRef]
31. Turki, B.M.; Parker, E.A.; Ziai, M.A.; Batchelor, J.C.; Sanchez-Romaguera, V.; Yeates, S.G. Study of printing errors in digitally fabricated FSS. In Proceedings of the 2013 Loughborough Antennas & Propagation Conference (LAPC), Loughborough, UK, 11–12 November 2013; pp. 429–432.
32. Jin, J.; Lin, Y.; Song, M.; Gui, C.; Leesirisan, S. Enhancing the electrical conductivity of polymer composites. *Eur. Polym. J.* **2013**, *49*, 1066–1072. [CrossRef]
33. Steurer, P.; Wissert, R.; Thomann, R.; Mülhaupt, R. Functionalized graphenes and thermoplastic nanocomposites based upon expanded graphite oxide. *Macromol. Rapid Commun.* **2009**, *30*, 316–327. [CrossRef]
34. Kapoor, A.; Mishra, R.; Kumar, P. Complementary frequency selective surface pair-based intelligent spatial filters for 5G wireless systems. *J. Intell. Syst.* **2021**, *30*, 1054–1069. [CrossRef]
35. Rahzaani, M.; Dadashzadeh, G.; Khorshidi, M. New technique for designing wideband one layer frequency selective surface in X-band with stable angular response. *Microw. Opt. Technol. Lett.* **2018**, *60*, 2133–2139. [CrossRef]

sensors

MDPI

Article

Planar Thermoelectric Microgenerators in Application to Power RFID Tags

Andrzej Dziedzic [1], Szymon Wójcik [1,*], Mirosław Gierczak [1], Slavko Bernik [2], Nana Brguljan [2], Kathrin Reinhardt [3] and Stefan Körner [3]

1 Wrocław University of Science and Technology, Wybrzeże Wyspiańskiego 27, 50-370 Wrocław, Poland; andrzej.dziedzic@pwr.edu.pl (A.D.)

2 Jožef Stefan Institute, Jamova cesta 39, 1000 Ljubljana, Slovenia

3 Fraunhofer IKTS, Winterbergstraße 28, 01277 Dresden, Germany; stefan.koerner@ikts.fraunhofer.de (S.K.)

* Correspondence: szymon.wojcik@pwr.edu.pl; Tel.: +48-791-425-025

Abstract: This paper presents an innovative approach to the integration of thermoelectric microgenerators (μTEGs) based on thick-film thermopiles of planar constantan–silver (CuNi-Ag) and calcium cobaltite oxide–silver ($Ca_3Co_4O_9$-Ag) thick-film thermopiles with radio frequency identification (RFID) technology. The goal was to consider using the TEG for an active or semi-passive RFID tag. The proposed implementation would allow the communication distance to be increased or even operated without changing batteries. This article discusses the principles of planar thermoelectric microgenerators (μTEGs), focusing on their ability to convert the temperature difference into electrical energy. The concept of integration with active or semi-passive tags is presented, as well as the results of energy efficiency tests, considering various environmental conditions. On the basis of the measurements, the parameters of thermopiles consisting of more thermocouples were simulated to provide the required voltage and power for cooperation with RFID tags. The conclusions of the research indicate promising prospects for the integration of planar thermoelectric microgenerators with RFID technology, opening the way to more sustainable and efficient monitoring and identification systems. Our work provides the theoretical basis and practical experimental data for the further development and implementation of this innovative technology.

Keywords: energy harvesting; thermoelectricity; thick film; thermocouple; thermopile; thermoelectric microgenerator; RFID tag

1. Introduction

With the rapid development of technology and the increasing demand for modern and sustainable energy sources, scientists and engineers are paying increasing attention to innovative solutions in the field of thermoelectric microgenerators (μTEGs). An area of research is the use of μTEGs in combination with radio frequency identification (RFID) technology. This combination not only opens up new prospects in the field of autonomous energy sources but also enables the development of advanced tracking and identification systems.

Thermoelectric microgenerators (μTEGs), based on the Seebeck phenomenon, allow the conversion of temperature difference into electrical energy. Using this phenomenon creates the possibility of powering small electronic devices such as sensors, transmitters, or even RFID tags. In the context of RFID applications, μTEGs open new horizons for autonomous power and secure identification.

RFID technology, which relies on the wireless transmission of data using radio waves, is widely used in various fields such as logistics, transportation, inventory management, and sensorics. However, traditional power sources for active RFID tags, such as batteries, can generate problematic environmental pollution and require regular replacement, which in turn can lead to loss of functionality and cost increase.

Combining thermoelectric microgenerator technology with RFID opens up a number of potential applications in various industries, mainly due to the specific requirements of this technology. μTEGs operate effectively in the presence of a temperature gradient, which opens up possibilities for a number of practical applications.

One potential application area is manufacturing. It is possible to use heated components inside production machines, which can generate large temperature gradients between the hot component and the environment. Another area is the automotive industry, where many components, such as the engine or exhaust pipe, reach high temperatures during the fuel combustion process. Using a microgenerator in such a context makes it possible to heat one side of the thermocouple with hot elements while cooling the other side as a result of the air movement generated during driving.

The possibilities associated with cooling one side of the microgenerator instead of heating it are also under consideration. In the medical field, where special freezers are used to store medical preparations, it is possible to imagine a situation where one side of the microgenerator is attached to human skin and the other side is exposed to extremely low temperatures as low as −30 °C.

The space industry also represents a promising market for applications of this technology, mainly due to low temperatures in space. One side of the microgenerator can be placed outside, where it is exposed to extreme conditions, while the other side can be placed in an environment tuned for human life, such as inside a space station or rocket.

This article presents the possible use of planar μTEGs in the context of powering RFID tags (active or semi-passive). Verification of the benefits of this relationship, analysis of technological possibilities, and prospects for further development in this area will be undertaken.

The main purpose of the analysis was to explore the potential of combining active or semi-passive RFID tags with planar thermoelectric microgenerators (μTEGs) due to their characteristics. RFID tags with an additional power supply offer significantly better communication performance, such as signal strength, resulting in a longer communication range. With appropriate μTEGs, it is possible to consider situations in which the RFID tag is self-sustaining and does not require additional intervention throughout the life of the device while generating a signal strength comparable to tags powered by traditional sources such as batteries.

Due to the distinctive characteristics of μTEGs, they allow the extraction of energy from the environment, converting thermal energy into electrical energy. This ability means that the microgenerators can be used effectively only where there is a temperature gradient. This results in the fact that the implementation under consideration cannot be applied universally. The main area of application is where there is a natural temperature gradient. Instead of wasting this energy, it can be efficiently used to power active or semi-passive RFID tags. Due to the specific requirements of μTEGs, they do not compete directly with other types of energy conversion, such as solar, mechanical, or RF. They are a substitute that expands the range of power options. The final choice between thermal, solar, or mechanical energy conversion and traditional battery systems to power RFID tags depends on the environmental conditions at the site.

This paper attempts to simulate planar thermoelectric microgenerators in various configurations. Based on tests conducted for actual microgenerators, simulations were performed with structural modifications to achieve the proper thermoelectric strength and power necessary for efficient operation in combination with a potential RFID tag. The objective of the analysis was to determine whether such a combination is possible.

To achieve this, in the first stage, two material systems—$Ca_3Co_4O_9$-Ag and CuNi-Ag—were experimentally studied and used to fabricate planar thermoelectric microgenerators in thick-film technology. Based on the measurements of the thermoelectric force generated by them and changes in their internal resistance as a function of the temperature difference between cold and hot thermoelectric junctions, this is as a function of the temperature gradient, and the parameters of individual thermocouples were determined, i.e., the characteristic dependence of their Seebeck coefficient and internal resistance (and thus resistivity) on temperature.

In the next step, the values of the thermoelectric force and internal resistance of planar μTEGs were simulated depending on the number of thermocouples in the thermopile, the geometric dimensions of the arms of a single thermocouple, the reference temperature, and the value of the temperature gradient. The obtained results were compared with the energy requirements of the RFID tags, which led to the conclusion that the proposed solutions could find applications to power active or semi-active RFID tags.

In the future, the practical implementation of this solution is planned, which, according to simulations, should work properly under the relevant conditions defined in this paper.

2. Thermoelectric Microgenerators (μTEGs)

μTEGs are devices that convert the temperature difference between two junctions of conductors or semiconductors into electrical energy using the Seebeck effect. This type of technology is widely used and is gaining popularity due to its features [1]. The main component of a microgenerator is the thermoelectric module, also called the thermopile, which consists of many pairs of thermocouples. The most common materials in microgenerators are usually bismuth, tellurium, and lead-based compounds [2]. These materials have thermoelectric properties that allow them to efficiently convert the temperature difference into electrical energy.

Thermoelectric structures have different parameters due to the materials used. To verify the performance of the structures, appropriate correlations are used, allowing the different parameters to depend on each other. The most common parameters are the power factor (PF), figure of merit (Z), and the dimensionless figure of merit (ZT).

$$\mathrm{PF} = \frac{\alpha^2}{\rho}\left[\frac{W}{m \times K^2}\right] \quad (1)$$

$$\mathrm{Z} = \frac{\alpha^2}{\rho \times \lambda} = \frac{\mathrm{PF}}{\lambda}\left[\frac{1}{K}\right] \quad (2)$$

$$\mathrm{ZT} = \frac{\alpha^2}{\rho \times \lambda} \times \mathrm{T} = \frac{\mathrm{PF}}{\lambda} \times \mathrm{T} = \mathrm{Z} \times \mathrm{T}[-] \quad (3)$$

where α is the Seebeck coefficient, ρ is the resistivity, λ is the thermal conductivity, and T is the temperature.

We can distinguish two main modes of operation of microgenerators, constant and variable. In the first case, there is a constant temperature gradient between junctions, and then the microgenerator generates a constant value of thermoelectric force and thus a constant electrical power. In the variable mode, on the other hand, the TEG generates a variable value of electrical power due to random or periodic changes in the temperature gradient. This is because the ambient temperature may not be constant and thus the voltage generated by the structure is variable. When selecting a microgenerator, it is necessary to take into account the minimum and maximum temperatures between the junctions and make sure that it will be sufficiently effective in extreme cases. Otherwise, the microgenerator may provide less energy than expected, and, as a result, the system will not work properly without another power source.

μTEGs have found applications in powering low-voltage devices such as sensors, clocks, and small electronic portable devices [3]. They are used in health monitoring systems, where they can power various actuator components [4]. A receiving group is also the automotive industry, where thermoelectric microgenerators can convert the heat generated during the operation of an engine or other components into additional electrical energy [3]. In the industry, μTEGs can be used to generate electricity where there are temperature differences between production processes.

Despite their many advantages, μTEGs face challenges, such as improving conversion efficiency, reducing manufacturing costs, and developing more efficient thermoelectric materials. However, with continued advances in science and technology, μTEGs have the potential to become an important component in the field of sustainable energy and portable electronics [5].

3. RFID Tags

RFID technology allows for contactless identification of objects using radio waves. An RFID system consists of two main components: a reader (or readers) and RFID tags [6,7]. The RFID reader and the RFID tag are the two key components of RFID-based identification systems. These two elements work together to create an efficient way to track, identify, and manage data. However, there is a fundamental difference between them, both in terms of function and the role they play in the system.

An RFID reader is a device that has the ability to receive and interpret radio signals transmitted by RFID tags. It acts as the interface between the tag and the data management system. The reader is usually a powered device that actively sends radio signals toward the tags and receives responses while identifying each tag's unique identification number. RFID readers are widely used in various fields, such as logistics, warehousing, transportation, and access control systems [6,7].

An RFID tag is a small and usually inexpensive device that contains an antenna and a microchip or other form of memory. It stores information, such as an identification number or data about the product or object to which the tag is attached. RFID tags often do not have their own power source as they are activated by energy from radio signals sent by the RFID reader when approached—these are passive tags. There are also active tags with their own power supply or semi-passive tags that combine the functionality of active and passive tags. There are many different types of RFID tags, including those designed for one-time use and more advanced reusable ones [6,7].

Passive tags are often cheaper, smaller, and lighter, making them ideal for applications where size constraints are important. Passive tags can be used in areas such as product marking, inventory management, logistics, or access control. They can also work with various types of sensors, but a power supply is often required. Active RFID tags, unlike passive tags, have their own power source (usually a battery). As a result, they can transmit data over longer distances and offer additional functions. They are usually larger and more expensive. They are used, for example, in vehicle tracking, real-time monitoring, or access control systems. Semi-passive tags have their own power source but also use energy from the reader during communication. They represent a compromise between the range of active tags and the smaller size of passive tags. Their application area is a combination of applications for active and passive tags [8–11].

Communication between the reader and the tag can follow two basic communication procedures called full/half duplex (FDX/HDX) or sequential (SEQ). Both communication methods can be used for all types of tags. In multidirectional mode (full duplex mode), the tag and the reader can transmit data simultaneously in both directions. This means full two-way communication. To support this mode, a continuous power supply to the tag is required for the period of communication. In the single-directional mode (half duplex), communication between the tag and the reader is alternate. Only one device can transmit data at a time. A continuous power supply to the tag for the period of communication is required to support this mode. In sequential mode, communication between the tag and the reader is alternated. At any given moment in time, only one device transmits information. The sequential mode is characterized by the fact that only a temporary power supply to the tag is required [12].

With different types of RFID tags and modes of operation, the technology is used in a wide range of fields, from logistics to security systems, bringing significant improvements in the identification, tracking, and management of various types of objects.

4. Powering RFID Tags—A Short Literature Review

The planned integration of μTEG with RFID tags requires the determination of the electrical parameters necessary for the proper operation of the system. Based on the literature, it can be observed that the minimum voltage needed to power the microprocessor is in the range from 1 V to 1.5 V [13–16]. An RFID chip powered by 1.5 V, which is mainly due to the processor used is described in [13]. Due to the widespread use of 1.5 V batteries, many low-voltage systems are based on this voltage. This trend is also described in [15], where a special processor has been designed for 1.5 V so as to achieve low current consumption. The main reason why they operate at voltages above 1 V is the processor [13,16]. At this very moment, circuits operating at low voltages are being researched, which is also shown in [16], where a special processor for RFID tags that can operate from 1 to 1.8 V is presented. For example, it is shown that the RFID tag works correctly with a power supply voltage of 1.2 V [14].

However, it should be noted that the basic parameter that defines the operation of the system is the power. The minimum power defined in the literature for the basic operation of an RFID tag is 8 μW. Such a low power is already sufficient to read basic information such as ID [14]. With specially designed systems this value can be even lower; the literature reports that in extreme conditions it is even possible to operate at a power of 3.15 μW [15]. However, it is generally assumed that the commonly accepted range of power is from 8 μW to 1 mW [7,14]. Note that when calculating power, it must be taken into account that the power generated by the energy source is not the final power. Assuming resistance matching of the generator and the load, ¼ of the generator power should be assumed [13].

Due to the application, various sources of renewable energy are used to power RFID tags. The basic principle for all types of conversion is to acquire ambient energy and then, using appropriate conversion circuits, accumulate it together with RF energy. This solution allows for proper functioning, ensuring continuous operation. Applications are described, for example, in [17], which focuses on an overview of solutions. Photovoltaic conversion is suitable for general applications, especially for supply chains, construction sites, and industry. RF conversion is also suitable for general applications, medicine, and agriculture. In addition, the area of interest is thermal conversion (using TEG), which can be excellent for industrial and medical care [17]. Of course, application areas can be very broad for all methods. The most important is the general scheme for connecting an RFID tag with a harvester chip. Due to the processors used, it is very important to provide a constant and relatively high voltage. For this purpose, suitable step-up converters should be used, which provide the required electrical parameters to power the processor chip and the RFID tag interface.

Very often, unconventional energy sources are proposed to power RFID tags. This is connected with problems during the use of RFID tag sensors, which relied mainly on batteries. For example, an RFID system powered by a small, flexible photovoltaic panel is described in [18]. During real-world measurements, the indoor reading range was up to 23 m. The entire system required about 168.3 μW. Of course, the processed energy must meet minimum power criteria. The energy collected thus is then stored using, for example, capacitors or batteries [10]. In the end, the harvested energy can be a component for powering the RFID tag.

Piezoelectric energy conversion is also commonly considered. A suitable generator can convert the shock energy of various naturally occurring mechanical vibrations into electrical energy. In [19], the authors attempted to analyze the combination of RFID tags and a piezoelectric generator due to the need for periodic battery replacement. The authors elaborate on possible applications of piezoelectric generators in active RFID tags. As examples of applications of RFID tags with piezoelectric generators, they proposed, for example, a security system to monitor children going to and returning from school and a network system to monitor the use of a conference room [19]. A microvibrating MEMS system may also find its application [20]. The energy provided by the vibrating system is stored in a capacitor and released by a mechanical switch to a tag. During tests, communication of up to 15 m was possible when the chip was powered by a microvibrator [20].

The literature also describes combinations of thermoelectric technology with RFID tags. The integration of an RFID tag with a flexible thermoelectric generator (TEG) is presented in [21]. It consists of an EPC C1G2/ISO 18000-6C ultrahigh-frequency RFID integrated circuit connected to a low-power microcontroller unit [22]. With a gradient of 2.5 °C, the TEG generates an output power of 400 μW at an output voltage of 40 mV. Through the use of a step-up converter, the appropriate electrical values are supplied to the RFID tag. During semi-passive operation, a communication range of up to 22.2 m has been measured [21]. It is also possible to combine the antennas with a thermoelectric generator [23]. The TEG can be mounted on top of the antenna, and its main task is to convert a thermal gradient between the antenna and the environment. The antenna is properly designed to ensure good heat transfer between the antenna and the TEG [23].

An interesting application is the design of a battery-free wearable system that measures the skin temperature of the human body while collecting energy from body heat using TEG. The system consists of a UHF RFID temperature sensor placed on the patient's arm, supported by additional power from an energy harvesting module that collects the thermal energy emitted by the patient's body. The experimental results of thermal energy harvesting are presented, and the module is characterized under various conditions, such as immobility and indoor and outdoor walking [24]. Finally, the tag was tested under fully passive and assisted operating conditions. The results show that the communication range of the RFID sensor is improved by 100% when taking measurements every 750 ms and by 75% when taking measurements every 1000 ms when the sensor is assisted by the energy harvesting module.

A hybrid system that combines radio frequency (RF) signals and energy harvested thanks to thermoelectric phenomena was also proposed. The main objective was to combine energy from both sources to solve the limited lifetime of semiactive UHF RFID tags, which are battery-powered due to the specific requirements of their applications. The simulation results show that the system can operate even at very low input voltages, such as 31.62 mV, and is capable of harvesting energy from a single source or both sources simultaneously. From the simulations, the maximum output voltage of 3.44 V was obtained from a single RF source of 0 dBm. A maximum output voltage of 3.40 V was obtained when energy was collected from both sources, RF and thermoelectric, simultaneously, with a power of 0 dBm and a temperature difference of 10 K [25].

On the basis of the literature, we see that TEG technology is being considered in combination with RFID tags. Our most important goal is to verify whether specific materials, $Ca_3Co_4O_9$ and CuNi, have the potential to be used as energy sources for RFID tags. We plan to use a standard step-up converter to increase the voltage to the required level. In contrast to other publications, our application is based on a planar thermoelectric microgenerator (μTEG) instead of a generator (TEG). This results in a much smaller size (volume), which helps to ensure the compactness of the device. It is worth noting that due to the very small dimensions of commonly available RFID tags, microgenerators allow the final device to maintain very small dimensions. The general trend toward miniaturization also works in favor of the presented solution, and the simulations themselves are preliminary for further research.

5. Planar Thermoelectric Microgenerators (μTEGs)—Own Research

The research conducted was carried out using planar μTEGs (Figures 1 and 2). A key feature of planar TEGs is their flat structure, which means that they are built on a single plane. The flat structure of μTEGs has important applications in microsystems where small size is key. Due to their compact design, the systems are finding applications in fields such as microscale electronics, sensorics, medicine, or other areas where there is a need for an efficient, small, and self-sustaining energy source. Planar structures make it possible to power microelectronics in places that are hard to reach or provide autonomous energy for microsystems, eliminating the need for an external power source. Planar thermoelectric microgenerators thus represent a step forward in the development of microelectronics technology, opening up new perspectives for applications that require small and self-sustaining power sources. Planar structures consist of relatively thin thermoelectric layers placed on a flat substrate. The material is spread in a single plane, causing heat to flow along the substrate [5,26–29].

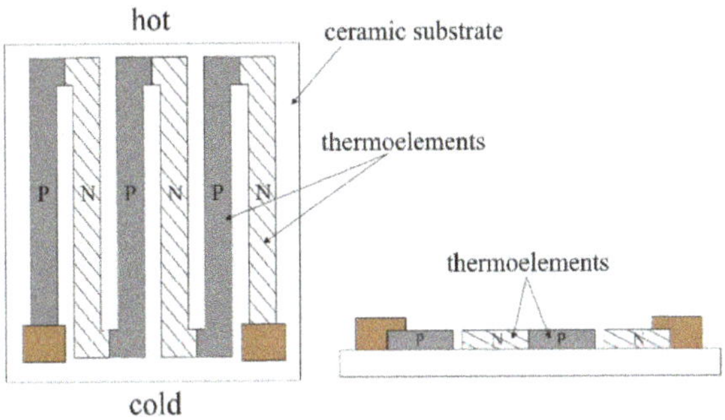

Figure 1. Schematic of a planar thermoelectric microgenerator (μTEGs).

Figure 2. Examples of planar μTEGs.

A resistive paste, whose main inorganic ingredient was constantan powder ($Cu_{55}Ni_{44}Mn$) [30], a paste prepared from powdered calcium cobaltite oxide ($Ca_3Co_4O_9$–Ca349) [31], and a commercial silver-based paste ESL 599E were used to fabricate thermopiles, which were composed of four thermocouples connected in series.

The test structures were realized using classic thick-film technology. For this purpose, photolithographic masks were made in the first step, and then screens were made; the mask and screen for one material are shown in Figure 3. Stainless screens with a density of 200 mesh were used. The next step was the screen printing of pastes on alumina substrates, for which a semi-automatic Uniprint PMGo3V screen printer was used. After applying, the first arms (pastes with the functional phase CuNi or $Ca_3Co_4O_9$ functional phase) were applied and the layers were dried at 120 °C for 10 min, after which a part of the structures was subjected to isostatic compression at 200 bar. Films prepared this way were fired in a six-zone tunnel furnace BTU QA-41-6-54 on a 60 min cycle with a 10 min hold at peak temperature. The second arms (from Ag-based ink) were then printed and, after drying, fired at 500 °C peak temperature. The CuNi-Ag structures were fired in a nitrogen atmosphere, while the $Ca_3Co_4O_9$-Ag thermopiles were fired in an air atmosphere. The basic technological steps are listed in Table 1.

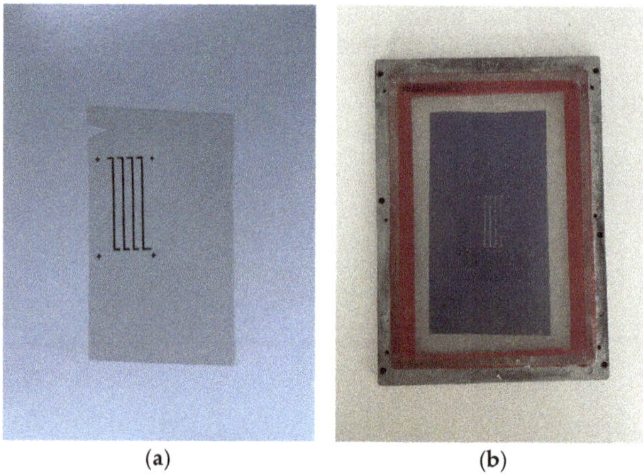

Figure 3. (a) Photolithography mask; (b) screen printing screen.

Table 1. The structures studied.

Material 1	Application Process—Material 1	Isostatic Compression before Firing—Material 1	Firing—Material 1 [°C]	Material 2	Application Process—Material 2	Isostatic Compression before Firing—Material 2	Firing—Material 2 [°C]
$Ca_3Co_4O_9$	Screen Printing	-	900	Ag	Screen Printing	-	500
$Ca_3Co_4O_9$	Screen Printing	200 bar	900	Ag	Screen Printing	-	500
CuNi	Screen Printing	-	850	Ag	Screen Printing	-	500
CuNi	Screen Printing	200 bar	850	Ag	Screen Printing	-	500

The parameters of the CuNi-Ag and $Ca_3Co_4O_9$-Ag microgenerators were determined over a wide temperature range. Measurements were carried out on a specially prepared measurement stand, the purpose of which was to properly heat/cool the structure and to perform point measurements of temperature, thermoelectric force, and internal resistance of microgenerators [32]. The stand is based on the cooperation of three modules: a desktop computer with a suitable application, a data logger, and a measuring table with a controller. After selecting the appropriate mode of operation, communication with the data logger and the controller is initiated. This enables correct operation and continuous data collection during measurements. Figure 4a shows the measuring station. Meanwhile, the arrangement of the measurement probes in the characterization of a planar thermoelectric microgenerator on an alumina substrate can be seen in Figure 4b.

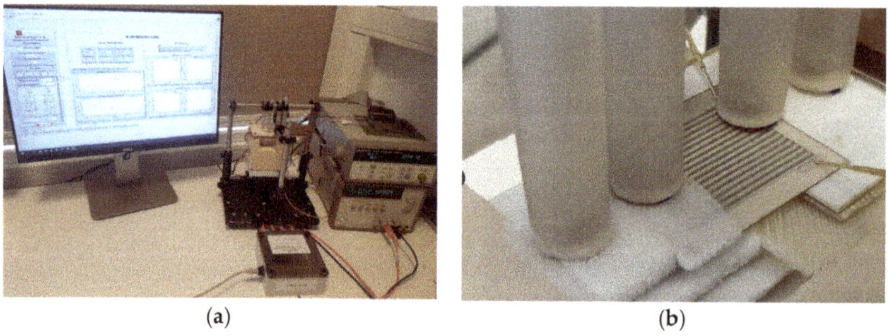

Figure 4. (a) Measurement stations for the characterization of thermoelectric microgenerators; (b) example of a characterization process.

Measurements of the thermoelectric properties of the test structures over a wide temperature range were carried out in two stages. In the first stage, one side of the µTEG structure is maintained at room temperature, while the other side is gradually heated to about 450 K using a thick film heater. In each successive step of 10 min, the heater power was increased by 10%. After the structure was cooled to room temperature, a stepwise cooling of the other side was performed to about 230 K using a Peltier module. In this stage, each step took approximately 7 min to stabilize the temperature on the cooled side of the structure. To determine the thermoelectric properties of the structures, the thermoelectric force and internal resistance values of the µTEGs measured under quasi-static conditions, that is, at the end of each heating or cooling step, were used.

Research made it possible to determine the Seebeck coefficient and the resistivity of the structure. From the measurements, the point Seebeck coefficient (small increment defined in Equation (4)) was determined, and the parameters were converted for a single thermocouple. The graphs (Figure 5) show the change in the Seebeck coefficient versus operating temperature for two different materials. The graphs presented show that the Seebeck coefficient is not a constant parameter, and that the operating temperature matters for the final performance of the thermoelectric microgenerator. Determining the Seebeck coefficient for a single thermocouple makes it possible to define how many thermocouples to apply to obtain the desired supply voltages.

$$S(T_n) = \frac{E(T_n) - E(T_{n-1})}{T_n - T_{n-1}} \tag{4}$$

where: $T_n \approx T_{n-1} + 10$ K.

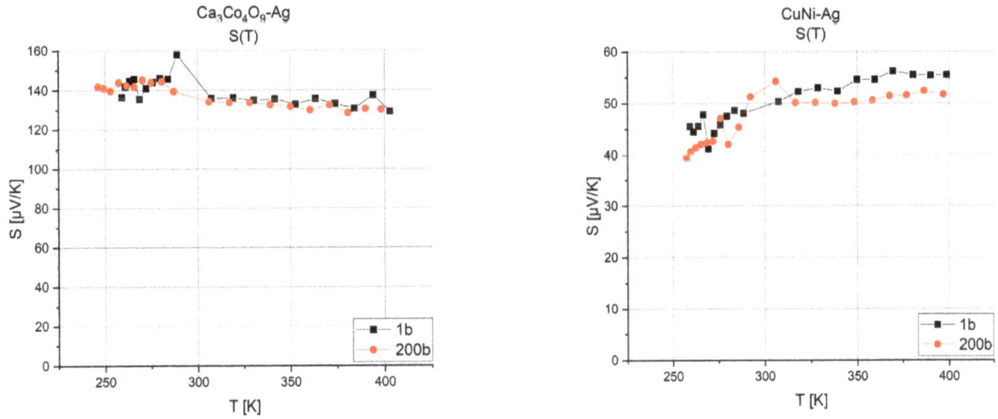

Figure 5. Seebeck coefficient versus temperature for compressed and uncompressed $Ca_3Co_4O_9$-Ag and CuNi-Ag structures.

On the basis of the value of the Seebeck coefficient and resistivity, the power factor (PF) was determined (Figure 6), which allows the determination of the power output of a single thermocouple. It allows the selection of a material with the best power properties at a given operating temperature. This is a very important parameter that includes the value of the Seebeck coefficient and the resistivity of the material.

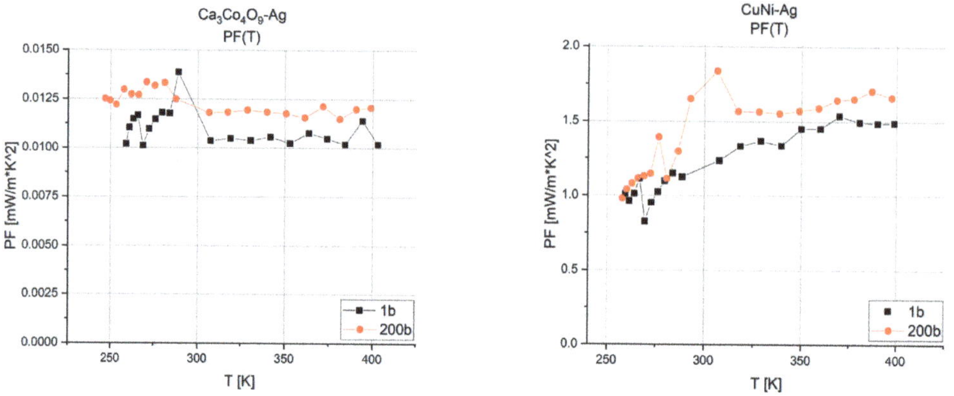

Figure 6. Power factor versus temperature for compressed and uncompressed $Ca_3Co_4O_9$-Ag and CuNi-Ag structures.

6. Combining μTEGs with RFID

Based on the literature, it can be observed that active RFID tags require a minimum voltage of 1 V to 1.5 V for proper operation [13–16] and a power level of 8 μW to 1 mW [7,13,14], depending on the type of tag. When appropriate measures are taken for the generation of a microgenerator, the resulting parameters will allow the system to work properly. As a result, the device will be able to communicate over a longer distance and will not rely only on the energy captured from radio waves. It is also important that the tag can operate in semi-passive mode, which is even the preferred method of cooperation between RFID tags and thermoelectric microgenerators. This significantly reduces power requirements, and the main task is to increase the communication distance [21,23–25]. Based on the results obtained for thermoelectric structures, CuNi- and $Ca_3Co_4O_9$-based, it can be concluded that a suitable combination of microgenerators together with RFID tags is possible. The resulting device can act as a semi-passive or active tag.

According to the measurement method described in Section 5, the heating ranges were selected for the $Ca_3Co_4O_9$-Ag- and CuNi-Ag-based structures (structures without isostatic compression), and then the corresponding simulations were performed to determine the parameters of the thermoelectric force and the power factor depending on the number of thermocouples and the resistance of the system.

The first possibility is to increase the number of thermocouples on the substrate. Figure 7 shows the voltage that can be obtained depending on the temperature gradient between the junctions for microgenerators consisting of 10, 50, 100, or 150 thermocouples.

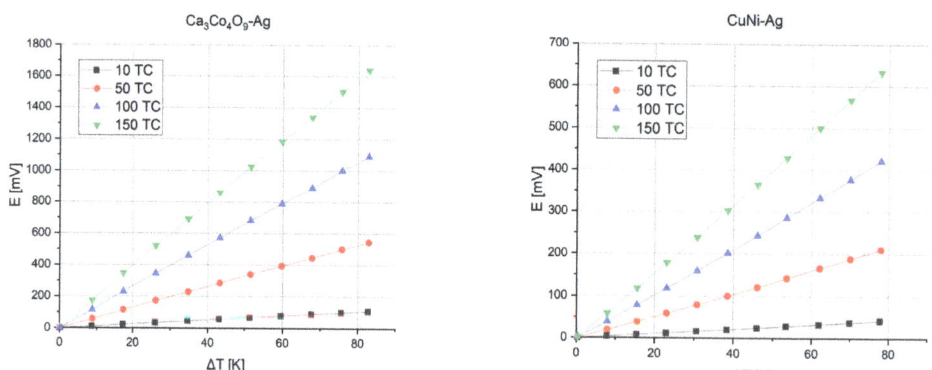

Figure 7. Thermoelectric force versus temperature gradient and number of thermocouples.

To calculate the power, it is necessary to take into account the change in resistance that occurs with a change in temperature. The resistance for a single thermocouple is presented in Table 2.

Table 2. Resistance of a single thermocouple versus the temperature gradient (cold junctions at room temperature).

ΔT [K]	0	8	16	24	32	40	48	56	64	72	80
$Ca_3Co_4O_9$-Ag-1b [kΩ]	7.079	7.032	6.983	6.930	6.878	6.818	6.763	6.704	6.643	6.578	6.486
CuNi-Ag-1b [Ω]	8.122	8.108	8.097	8.097	8.099	8.100	8.105	8.111	8.128	8.139	8.148

According to Ohm's law, on the basis of the resulting voltage and resistance, it is possible to determine the power generated by the microgenerator.

$$P = \frac{U^2}{R} \tag{5}$$

It is important to note that, for example, in the case of a DC circuit when matching due to the maximum power transferred from the generator to the load, which occurs when the resistance of the load is equal to the internal resistance of the source, the power given to the load is equal to ¼ of the power generated by the thermoelectric microgenerator.

Figure 8 presents the power obtained from such structures. As the temperature gradient increases, the voltage and, as a result, the power that the structure is able to generate also increase.

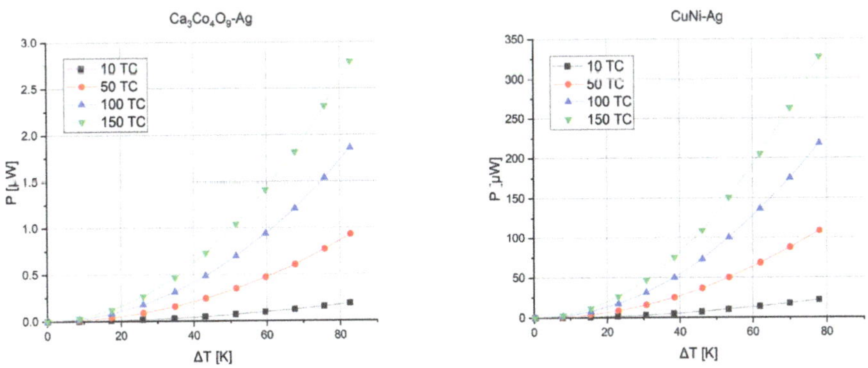

Figure 8. Generated power versus temperature gradient and number of thermocouples.

Another possibility is to reduce the resistance of a single thermocouple. This will not affect the generated thermoelectric force because the Seebeck coefficient is a material parameter. But, on the other hand, it will allow more current to flow. When the internal resistance of a single thermocouple is reduced, for example, by 50%, respectively, it is possible to obtain a higher system power.

For the test structure studied, the number of squares per thermocouple is about 118. When reducing this number to 59 (50%), the resistance of the structure will decrease proportionally, and thus more power will be generated (Figure 9).

By combining the two methods, it is possible to obtain the appropriate voltage and power required to power RFID tags in active or semi-passive mode. This is a promising application for ensuring an appropriate temperature gradient between the junctions and optimizing the thermoelectric microgenerator.

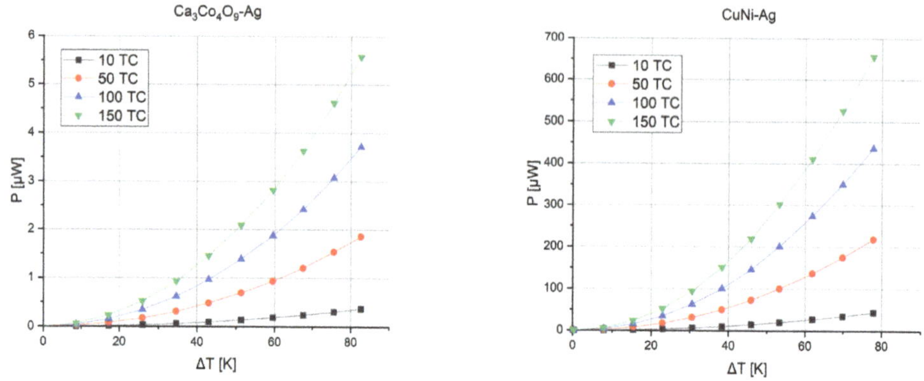

Figure 9. The generated power versus the temperature gradient and the number of thermocouples for resistance reduced by 50%.

On the basis of simulations, it was concluded that, given the relatively high power of the CuNi-Ag structure, it is appropriate to carefully consider the application of this microgenerator. In further analysis, a 10 × 50 mm^2 substrate was adopted and simulations were carried out for four different thermocouple sizes. In the calculations sheet resistance, R□ = 0.060 [Ω/□] was taken for CuNi- and R□ = 0.012 [Ω/□] for Ag-based thick films. During the calculations, the sizes of the paths made of constantan and silver were modified. When determining the test cases, the possibilities associated with thick-film technology were taken into account. A schematic of the simulated structure is shown in Figure 10.

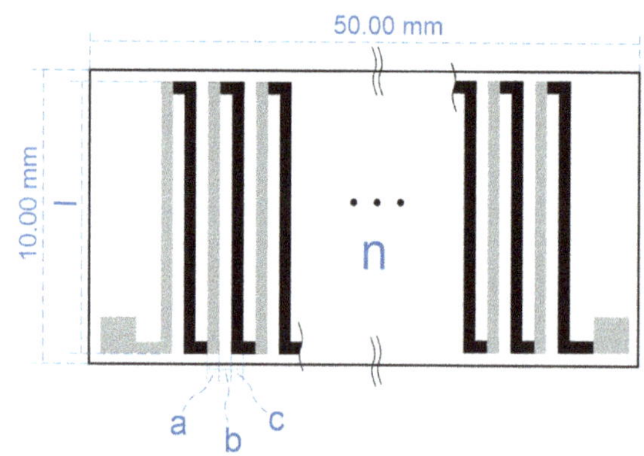

Figure 10. Schematic of a microgenerator on a substrate of size 10 × 50 mm^2, where the following holds: a—width of the silver path; b—gap between paths; c—width of the constantan path; l—length of the single path; n—number of thermocouples.

The four simulated structures are shown in Table 3, while the corresponding resistances are shown in Table 4. Structure 4 was specifically designed so that the resistances of the two paths were similar.

Table 3. Simulated structures with realistic dimensions.

Structure No.	a [mm]	b [mm]	c [mm]	l [mm]	n_{max} [-]
1	0.2	0.2	0.2	9	60
2	1	0.2	1	9	20
3	1	0.2	0.2	9	30
4	0.2	0.2	1	9	30

Table 4. Resistance of the simulated structures at the maximum number of thermocouples.

Structure No.	R_{CuNi} [Ω]	R_{Ag} [Ω]	$R_{CuNi} + R_{Ag}$ [Ω]	Relationship
1	176.4	32.4	208.8	$R_{CuNi} > R_{Ag}$
2	13.68	2.16	15.84	$R_{CuNi} > R_{Ag}$
3	102.6	3.24	105.84	$R_{CuNi} > R_{Ag}$
4	17.64	16.2	33.84	$R_{CuNi} \approx R_{Ag}$

On the basis of the specified parameters, the thermoelectric force and power of the designed microgenerators were calculated. Figure 11 presents the voltages and powers possible from the test microgenerators.

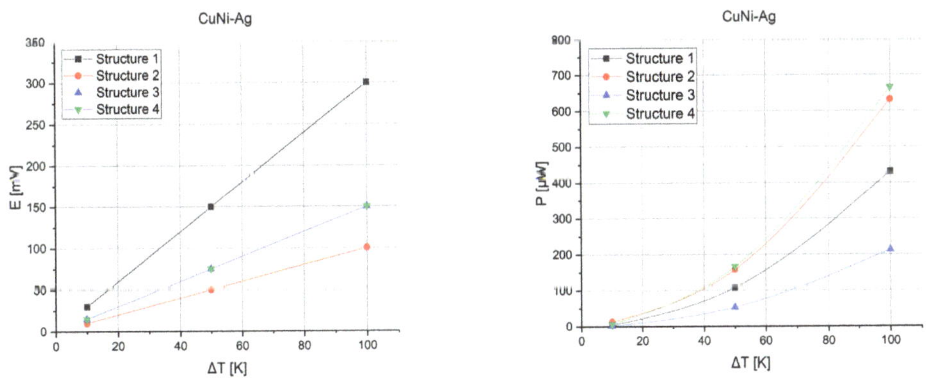

Figure 11. Voltage and power for four simulated structures at three gradient points of 10 K, 50 K, and 100 K.

The determined characteristics show that the best power performance is generated by the fourth structure, reaching about 170 µW at a 50 K gradient and about 670 µW at a 100 K gradient. On the basis of these results, it can be concluded that the microgenerator is suitable for use with RFID tags using a suitable step-up converter. During conversion with resistance matching, it must be assumed that the effective power will be lower according to relation 6. Figure 12 presents a schematic connection of the circuit to ensure correct operation. Table 5 shows the powers obtained with conversion for converters with 90% efficiency.

$$P = \frac{E^2}{4 \cdot R} \quad (6)$$

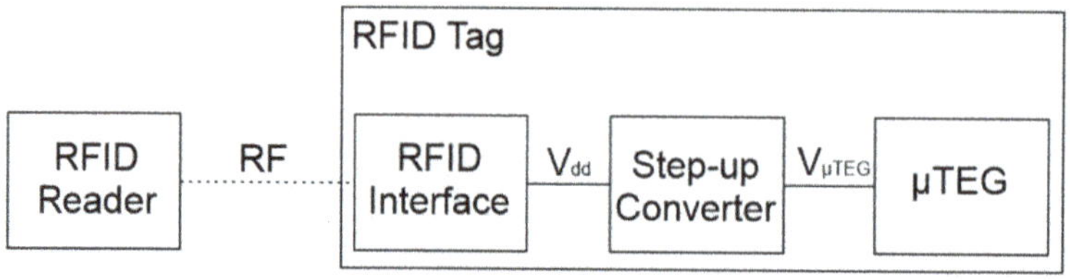

Figure 12. Schematic of μTEG's connection to RFID.

Table 5. Comparison of voltage and power values before and after conversion for the fourth structure.

ΔT [K]	U—before Conversion [mV]	$P_{\mu TEG}$—Effective with Load [μW]	U—after Conversion [mV]	$P_{converter}$—after Conversion Where η = 90% [μW]
10	15	1.66	1200	1.50
20	30	6.65	1200	5.98
30	45	14.96	1200	13.46
40	60	26.60	1200	23.93
50	75	41.56	1200	37.40
60	90	59.84	1200	53.86
70	105	81.45	1200	73.30
80	120	106.38	1200	95.74
90	135	134.64	1200	121.18
100	150	166.22	1200	149.60

According to the assumptions, it can be seen that with a voltage boost to 1.2 V, the microgenerator can provide adequate voltage and power for the RFID tag with resistance matching. It should be noted that a minimum power of 8 μW [14] has already been obtained with a gradient of less than 30 K (Figure 13).

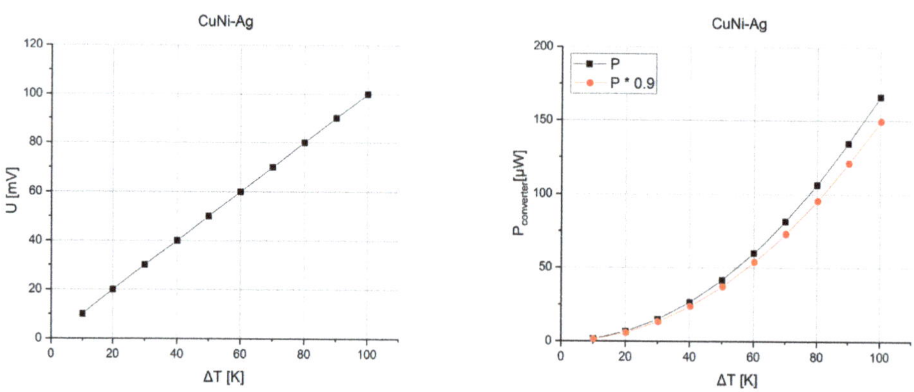

Figure 13. Voltage before conversion and effective power (100%, 90%) for the fourth structure.

By analyzing the presented structure, it is possible to show which gradient is necessary for the operation of the RFID tag per mm^2 of the entire structure, starting from a gradient of 30 K since, as described above, this is considered the minimum required gradient (Table 6).

Table 6. Power distribution per mm^2 of structure, for the fourth structure with dimensions of 10 × 50 mm^2.

ΔT [K]	30	40	50	60	70	80	90	100
P/A [μW/mm^2]	0.0299	0.0532	0.0831	0.1197	0.1629	0.2128	0.2693	0.3324

Based on the sources, the required energy per bit needed for specific communication protocols was estimated. According to the article [33], the approximate (averaged) energy was determined for the RTF, TTF, and 808.15.4 protocols. By approximating the results, the values presented in Table 7 were obtained.

Table 7. Energy estimated per bit relative to the protocol.

Protocol	RTF	TTF	808.15.4
E_b [nJ/bit]	315.91	448.16	12.33

Knowing the energy per bit, the number of bits that can be sent using a particular communication protocol was determined. Since the 808.15.4 protocol requires much less energy to transmit data, it allows a much higher number of bits to be sent. Figure 14 presents the frequency at which bits can be sent, depending on the thermal gradient.

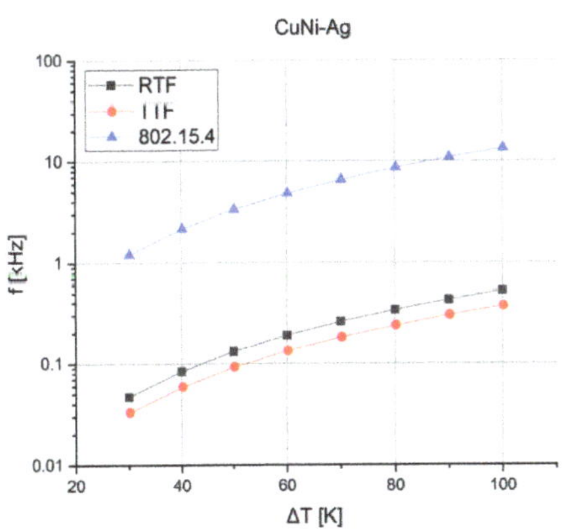

Figure 14. Bit transfer depends on the protocol and on the μTEG thermal gradient for the fourth structure.

With a one-second operating period assumed, the energy that the structure can generate depending on the gradient was determined. The results are summarized in Table 8.

Table 8. Energy distribution per mm^2 of the structure, for the fourth structure with dimensions of 10 × 50 mm^2.

ΔT [K]	30	40	50	60	70	80	90	100
E/A [nJ/mm^2]	29.920	53.191	83.112	119.681	162.899	212.766	269.282	332.447

7. Conclusions

This article presents an innovative approach to powering RFID tags through the use of planar μTEGs. Research confirms that these microgenerators can provide an efficient and sustainable power source for RFID tags, opening up new possibilities in the field of identification and monitoring technology.

The main conclusion of the research is that planar μTEGs can work well with RFID tags, enabling not only their continuous and autonomous operation but also eliminating the need for traditional power sources such as batteries. This results in increased reliability and durability of RFID-based identification systems.

According to the results, the $Ca_3Co_4O_9$-based microgenerator only partially meets the criteria for working with active RFID tags. Despite the relatively high voltage generated by the $Ca_3Co_4O_9$-Ag structure, its power is low, which is due to its high internal resistance. A structure providing adequate performance would have to operate with a very large temperature gradient or consist of more than 300–500 thermocouples, which can be very difficult technologically. With 150 thermocouples and a temperature gradient of 60 K, a voltage of 1.2 V is possible, which is a desirable result. However, the final power is only 1.5 μW, which is more than five times less than the minimum required value, which is 8 μW.

CuNi-based microgenerators show much higher output power; however, their voltage is below the minimum level required for an active RFID tag. For this reason, these structures were carefully studied for their promising power performance when planning their physical implementation. Four different types of structure were simulated on a 10×50 mm^2 substrate. The best electrical performance was shown by a structure in which the width of the paths for silver was 0.2 mm and for constantan 1 mm, with a gap between them of 0.2 mm and a length of 9 mm. A maximum of 30 thermocouples fit on this structure, and all dimensions were adjusted according to the capabilities of thick film technology.

Simulations have shown that the CuNi-Ag structure is capable of working with RFID tags using a suitable boost converter that raises the voltage to 1.2 V. With a temperature gradient of 30 K and using a step-up converter with resistance similar to μTEG, 14.96 μW can be generated, suggesting the possibility of a practical application, assuming the minimum gradient is maintained.

Since different communication protocols require different amounts of energy for the transfer of a single data bit, an analysis of the frequency at which the tag can potentially send data bits depending on the power provided by the μTEG was conducted. According to the results, with a small thermal gradient (30 K), the microgenerator allows sending bits with suitable frequencies for RTF (~47 Hz), TTF (~33 Hz), and 802.15.4 (~1.2 kHz). As the thermal gradient increases, the effective power of the microgenerator also increases, resulting in a significant increase in frequency for the RTF (~0.53 kHz), TTF (~0.37 kHz), and 802.15.4 (~13.5 kHz) protocols at a gradient of 100 K.

From the results of this study, it can be concluded that the CuNi-Ag microgenerator can be effectively used to work with active or semi-passive RFID tags. As a result, it will make it possible to increase the communication distance and ensure continuous operation even in relatively low-temperature gradients.

On the basis of the research carried out, the practical implementation of the proposed solution of the thermoelectric microgenerator (μTEG) described above is planned.

In conclusion, μTEGs combined with RFID tags present themselves as a promising solution that not only increases the energy efficiency of identification systems, but also fits the global drive to create environmentally friendly technologies. The introduction of these innovative solutions can contribute significantly to the development of RFID technology towards a sustainable future.

Author Contributions: Conceptualization, A.D. and S.W.; methodology, A.D. and S.W.; investigation, S.W., M.G., S.B, N.B., K.R. and S.K.; validation, A.D., S.W., M.G., S.B., K.R. and S.K.; formal analysis, S.W. and A.D.; visualization, S.W., M.G. and A.D.; writing—original draft preparation, S.W.; writing—review and editing, S.W., A.D. and S.B.; supervision, A.D. All authors have read and agreed to the published version of the manuscript.

Funding: This research was funded by the statutory activity of the Faculty of Electronics, Photonics and Microsystems, Wroclaw University of Science and Technology.

Institutional Review Board Statement: Not applicable.

Informed Consent Statement: Not applicable.

Data Availability Statement: Data are contained within the article.

Conflicts of Interest: The authors declare no conflicts of interest.

References

1. Zulkepli, N.; Yunas, J.; Mohamed, M.A.; Hamzah, A.A. Review of Thermoelectric Generators at Low Operating Temperatures: Working Principles and Materials. *Micromachines* **2021**, *12*, 734. [CrossRef]
2. d'Angelo, M.; Galassi, C.; Lecis, N. Thermoelectric Materials and Applications: A Review. *Energies* **2023**, *16*, 6409. [CrossRef]
3. Ando Junior, O.H.; Maran, A.L.O.; Henao, N.C. A review of the development and applications of thermoelectric microgenerators for energy harvesting. *Renew. Sustain. Energy Rev.* **2018**, *91*, 376–393. [CrossRef]
4. Dargusch, M.; Liu, W.-D.; Chen, Z.-G. Thermoelectric Generators: Alternative Power Supply for Wearable Electrocardiographic Systems. *Adv. Sci.* **2020**, *7*, 2001362. [CrossRef]
5. Jaziri, N.; Boughamoura, A.; Müller, J.; Mezghani, B.; Tounsi, F.; Ismail, M. A comprehensive review of Thermoelectric Generators: Technologies and common applications. *Energy Rep.* **2020**, *6*, 264–287. [CrossRef]
6. Chetouane, F. An Overview on RFID Technology Instruction and Application. *IFAC-PapersOnLine* **2015**, *48*, 382–387. [CrossRef]
7. Want, R. An introduction to RFID technology. *IEEE Pervasive Comput.* **2006**, *5*, 25–33. [CrossRef]
8. Jankowski-Mihułowicz, P.; Kalita, W.; Skoczylas, M.; Węglarski, M. Modelling and design of HF RFID passive transponders with additional energy harvester. *Int. J. Antennas Propag.* **2013**, *10*, 242840. [CrossRef]
9. Gotfryd, M.; Pawłowicz, B.; Pitera, G. Active, passive and semipassive RFID systems. Series: Electrotechnics. *Sci. J. Rzesz. Univ. Technol.* **2015**, *34*, 243–258. [CrossRef]
10. Ferdous, R.M.; Reza, A.W.; Siddiqui, N.M.F. Renewable energy harvesting for wireless sensors using passive RFID tag technology: A review. *Renew. Sustain. Energy Rev.* **2016**, *58*, 1114–1128. [CrossRef]
11. Condemi, A.; Cucchiella, F.; Schettini, D. Circular Economy and E-Waste: An Opportunity from RFID TAGs. *Appl. Sci.* **2019**, *9*, 3422. [CrossRef]
12. Vyas, R.; Tye, B. A Sequential RFID System for Robust Communication with Underground Carbon Steel Pipes in Oil and Gas Applications. *Electronics* **2019**, *8*, 1374. [CrossRef]
13. Pillai, V.; Heinrich, H.; Dieska, D.; Nikitin, P.V.; Martinez, R.; Rao, K.V.S. An Ultra-Low-Power Long Range Battery/Passive RFID Tag for UHF and Microwave Bands with a Current Consumption of 700 nA at 1.5 V. *IEEE Trans. Circuits Syst. I Regul. Pap.* **2007**, *54*, 1500–1512. [CrossRef]
14. Astigarraga, A.; Lopez-Gasso, A.; Golpe, D.; Beriain, A.; Solar, H.; del Rio, D.; Berenguer, R. A 21 m Operation Range RFID Tag for "Pick to Light" Applications with a Photovoltaic Harvester. *Micromachines* **2020**, *11*, 1013. [CrossRef]
15. He, Y.; Hu, J.; Min, H. An ultra low-voltage, low-power baseband-processor for UHF RFID tag. *Front. Electr. Electron. Eng. China* **2008**, *3*, 99–104. [CrossRef]
16. Hu, J.; Wang, D.; Wu, J. A 2 Kbits low power eeprom for passive RFID tag IC. *Chin. J. Electron.* **2022**, *31*, 18–24. [CrossRef]
17. Solar, H.; Beriain, A.; Berenguer, R.; Sosa, J.; Montiel-Nelson, J.A. Semi-Passive UHF RFID Sensor Tags: A Comprehensive Review. *IEEE Access* **2023**, *11*, 135583–135599. [CrossRef]
18. Lemey, S.; Agneessens, S.; Van Torre, P.; Baes, K.; Vanfleteren, J.; Rogier, H. Wearable Flexible Lightweight Modular RFID Tag with Integrated Energy Harvester. *IEEE Trans. Microw. Theory Tech.* **2016**, *64*, 2304–2314. [CrossRef]
19. Takeuchi, M.; Matsuzawa, S.; Tairaku, K.; Takatsu, C. Piezoelectric Generaor as Power Supply for RFID-Tags and Applications. In Proceedings of the 2007 IEEE Ultrasonic Symposium, New York, NY, USA, 28–31 October 2007; pp. 2558–2561. [CrossRef]
20. Lu, Y.; Basset, P.; Laheurte, J.-M. Performance Evaluation of a Long-Range RFID Tag Powered by a Vibration Energy Harvester. *IEEE Antennas Wirel. Propag. Lett.* **2017**, *16*, 1832–1835. [CrossRef]
21. Solar, H.; Beriain, A.; Rezola, A.; del Rio, D.; Berenguer, R. A 22-m Operation Range Semi-Passive UHF RFID Sensor Tag with Flexible Thermoelectric Energy Harvester. *IEEE Sens. J.* **2022**, *22*, 19797–19808. [CrossRef]
22. *ISO/IEC 18000-6:2013*; Information Technology. Radio Frequency Identification for Item Management. Part 6: Parameters for Air Interface Communications at 860 MHz to 960 MHz. ISO: Geneva, Switzerland, 2013. Available online: https://www.iso.org/standard/46149.html (accessed on 20 February 2024).
23. Virili, M.; Georgiadis, A.; Niotaki, K.; Collado, A.; Alimenti, F.; Mezzanotte, P.; Roselli, L.; Carvalho, N.B. Design and Optimization of an Antenna with Thermo-Electric Generator (TEG) for Autonomous Wireless Nodes. In Proceedings of the IEEE RFID Technology and Applications Conference (RFID-TA), Tampere, Finland, 8–9 September 2014; pp. 21–25. [CrossRef]
24. Jauregi, I.; Solar, H.; Beriain, A.; Zalbide, I.; Jimenez, A.; Galarraga, I.; Berenguer, R. UHF RFID Temperature Sensor Assisted with Body-Heat Dissipation Energy Harvesting. *IEEE Sens. J.* **2017**, *17*, 1471–1478. [CrossRef]
25. Tengku Mohamad, T.N.; Sampe, J.; Berhanuddin, D.D. RF and thermal hybrid input for ultra-low power semi-active UHF RFID tags. In Proceedings of the 2018 IEEE 14th International Colloquium on Signal Processing & Its Applications (CSPA), Penang, Malaysia, 9–10 March 2018; pp. 47–51. [CrossRef]
26. Gierczak, M.; Prażmowska-Czajka, J.; Dziedzic, A. Thermoelectric Mixed Thick-/Thin Film Microgenerators Based on Constantan/Silver. *Materials* **2018**, *11*, 115. [CrossRef]

27. Markowski, P. Thermoelectric energy harvester fabricated in thick-film/LTCC technology. *Microelectron. Int.* **2014**, *31*, 176–185. [CrossRef]
28. Markowski, P.; Dziedzic, A. Planar and three-dimensional thick-film thermoelectric microgenerators. *Microelectron. Reliab.* **2008**, *48*, 890–896. [CrossRef]
29. Rudež, R.; Markowski, P.; Presecnik, M.; Košir, M.; Dziedzic, A.; Bernik, S. Development of thick-film thermoelectric microgenerators based on p-type $Ca_3Co_4O_9$ and n-type $(ZnO)_5{\cdot}In_2O_3$ legs. *Ceram. Int.* **2015**, *41*, 13201–13209. [CrossRef]
30. Sobek, P.; Reinhardt, K.; Kőrner, S. Reasonable resistor pastes with low TCR using constantan. In Proceedings of the 2021 23rd European Microelectronics and Packaging Conference & Exhibition (EMPC), Gothenburg, Sweden, 13–16 September 2021; p. 4.
31. Presečnik, M.; Bernik, S. The thermoelectric compound $Ca_3Co_4O_9$—Synthesis and characteristics. *AIP Conf. Proc.* **2012**, *1449*, 331–334. [CrossRef]
32. Wójcik, S.; Gierczak, M. System do oceny mikrogeneratorów termoelektrycznych w szerokim zakresie temperatury. *Przegląd Elektrotechniczny* **2022**, *98*, 131–135. [CrossRef]
33. Nilsson, B.; Bengtsson, L.; Bilstrup, U.; Wiberg, P.; Svensson, B. Towards an Energy Efficient Protocol for Active RFID. In Proceedings of the 2006 International Symposium on Industrial Embedded Systems, Antibes Juan-Les-Pins, France, 18–20 October 2006; pp. 1–4. [CrossRef]

sensors

MDPI

Article

Investigation of Factors Affecting the Performance of Textronic UHF RFID Transponders

Anna Ziobro [1,*], Piotr Jankowski-Mihułowicz [2,*], Mariusz Węglarski [2,*] and Patryk Pyt [2]

[1] Doctoral School of the Rzeszów University of Technology, 35-959 Rzeszów, Poland
[2] Department of Electronic and Telecommunications Systems, Rzeszów University of Technology, Wincentego Pola 2, 35-959 Rzeszów, Poland; p.pyt@prz.edu.pl
* Correspondence: d576@stud.prz.edu.pl (A.Z.); pjanko@prz.edu.pl (P.J.-M.); wmar@prz.edu.pl (M.W.)

Abstract: The aim of this paper is to demonstrate progress in textronic UHF RFID transponder (RFIDtex tag) technology. The fundamental idea behind the RFIDtex tag design involves galvanic separation between circuits of the sewn antenna and the chip, which are electromagnetically coupled through a system of inductive loops. To advance the development of this concept, it is crucial to detect factors affecting the performance of the transponders. To achieve this goal, a mathematical model of the textronic UHF RFID transponder was developed. It involves relationships that describe the impedance of each element, the mutual inductance of the loops, and the chip voltage, and it enables the exploration of the influence of these variables on general parameters such as impedance matching and read range. Various analytical and numerical approaches were considered to obtain the value of the mutual inductance of the loops. The dimensions and geometry of the antenna, as well as the matching circuit in the microelectronic module, were taken into account. Based on the mathematical model, it was determined that mutual inductance strongly affects the chip voltage for frequencies higher than 800 MHz. The calculations from the mathematical model were compared with numerical simulations. Experimental studies were also conducted to investigate how the transponder performance is affected by either the distance between the centers of the loops or the conductivity of the threads used to embroider the antenna. The measurement results allowed us to conclude that even small imperfections in the manufacturing of the transponder, which slightly increase the vertical or horizontal distance between the centers of the loops, cause a dramatic decrease in the mutual inductance and coupling coefficient, significantly impacting the transponder's performance.

Keywords: RFID transponder; smart textiles; smart fabrics; wearable devices; textile antennas; wearables; e-textiles; RFIDtex tag

1. Introduction

1.1. Textronic UHF RFID Transponders

In recent years, numerous research studies have addressed RFID wearables, mostly due to the expansion of the Internet of Things, as microwave and wireless systems are becoming eligible and ubiquitous in everyday life [1]. Among the developed solutions, a large portion pertains to medical and healthcare devices, such as sensors for measuring vital signals [1–3]. Other examples of RFID applications include positioning, localization, and tracking human activities [4,5]. Textile-based RFID tags need to meet various requirements like flexibility and low production costs, as well as challenges posed by mechanical stress, washing processes, or operation disruptions caused by proximity to the human body [6,7]. To achieve these goals, various conductive materials and technologies for their application are being investigated, such as nanocomposites, carbon nanotubes [8], conductive pastes, inks, fibers, or fabrics [9–13]. For example, according to the reference literature, graphene is used as a conductor due to its good mechanical and electrical properties. Its implementation in RFIDtex tag designs provides the opportunity to achieve a highly conductive, flexible, and mechanically robust material. Numerous examples of research studies in this area include conductive pastes [14], graphene sheets [15], laser-based methods such as laser-induced graphene [16–18], flexible graphene assembly films [19–21], or printing techniques [3,22–24]. However, the development and widespread use of these technologies encounter challenges, such as toxicity to humans and the environment, along with the need to implement a costly and complicated manufacturing process that is not commonly used or known in the textile industry [25]. Also, washability limitations of such experimental materials can be observed [26].

However, antennas in RFIDtex tags are frequently made by embroidering or sewing conductive threads [27–36], by knitting [37], or using any other method known in the textile industry. These proposals differ in their used threads, substrates, antenna structures, and types of integrated impedance matching elements. There are a wide variety of antenna types designed by researchers, including dipole, slotted patch, two symmetrical circular patches, meander line, fractal, rectangular or octagonal geometry, text-meandered, or even Mickey Mouse shaped. In the literature, alongside descriptions of works on specific solutions, there is also research aimed at solving problems that occur when embroidering, sewing, etc., like yarns deformation, accuracy of project reproduction, influence of machine settings, etc. [38–40]. Other papers address methods for determining the conductivity of embroidered antennas [35] or the use of polymers in conductive fibers [30,41], which are a desirable material in wearable electronics due to their flexibility [42–44].

Instead of using conductive fibers, other conductive flat materials can be employed in tag structures [45–48]. Such tags can be attached to the textile product using glue and ironing. In paper [48], a tag with a split-ring resonator antenna is proposed. It consists of two layers of conductive fabric. Nevertheless, other examples of UHF RFID tags made using commercially available conductive ink and the printing method can be found in references [12,49].

Among the solutions of wearable UHF RFID tags, there are also those that serve as sensors, including strain sensors [45], handwriting sensors [50], and surface crack monitoring [51]. On the other hand, the review in reference [52] addresses issues of e-textile UHF RFID tag design methodology, committed to overcoming problems related to interferences with their operation.

A particular problem for e-textile RFID transponders is the connection between the chip and the antenna. Efforts are undertaken to find ways to replace soldering, which has a temperature unsuitable for most textile materials [53]. One way is to use low-temperature soldering [26]. However, the most common way is applying bonding with conductive epoxy [27,31,35,47,48]. Also, examples of knitted [37] and embroidered interconnection can be found [29,54]. In paper [53], a comparison of three different methods of connection, sewing, snap buttons, and inserting, is presented. Also, a proposal for chipless embroidered RFID tags is found in reference [30]. Based on what has been presented so far, the work in reference [28] stands out, in that there is no physical connection between the antenna and the chip. The transponder consists of a linear dipole antenna and a rectangular coupling circuit of the chip. Both parts are embroidered and galvanically separated but inductively coupled. Changes in coupling between these components are used to measure displacement and strain.

This tag is similar to the textronic UHF RFID transponder [55], where the antenna and chip are inductively coupled instead of using a galvanic junction. This transponder differs from that in reference [28] in the construction of the antenna, where the coupling element is placed in the form of a loop, the method of fabricating the chip coupling circuit is different, and there is the application of both a linear dipole and a meander line antenna.

1.2. Aim of Research

This paper addresses issues related to textronic UHF RFID transponders which are integrated into fabrics. This idea is unique in terms of the galvanic separation between the transponder antenna and the chip. Due to the inductive coupling between these components, the RFIDtex tag construction includes two additional elements: a coupling circuit of the antenna and a coupling circuit of the chip. The antenna coupling circuit is fabricated using the same method as the antenna, whereas the chip coupling circuit, along with the galvanically connected chip, is made as a typical microelectronic device. Therefore, the concept envisions their production as separate modules: an antenna module and a microelectronic module. Both parts may be hidden in clothing, or they may also take the form of a decorative or functional garment element, such as a button, bead, emblem, etc.

In developing this concept, emphasis is placed on utilizing manufacturing techniques that are most widely used in the textile industry. For this reason, the antenna module is sewn or embroidered using conductive threads. Nevertheless, this concept allows for the use of other suitable materials and methods. The microelectronic module is manufactured separately and then distributed to clothing factories, where it can be easily sewn or attached to fabric. The numerical model of the textronic UHF RFID transponder's components and fabricated samples of two constructions that differ in the antenna type are presented in Figure 1. The microelectronic module is the same in both cases.

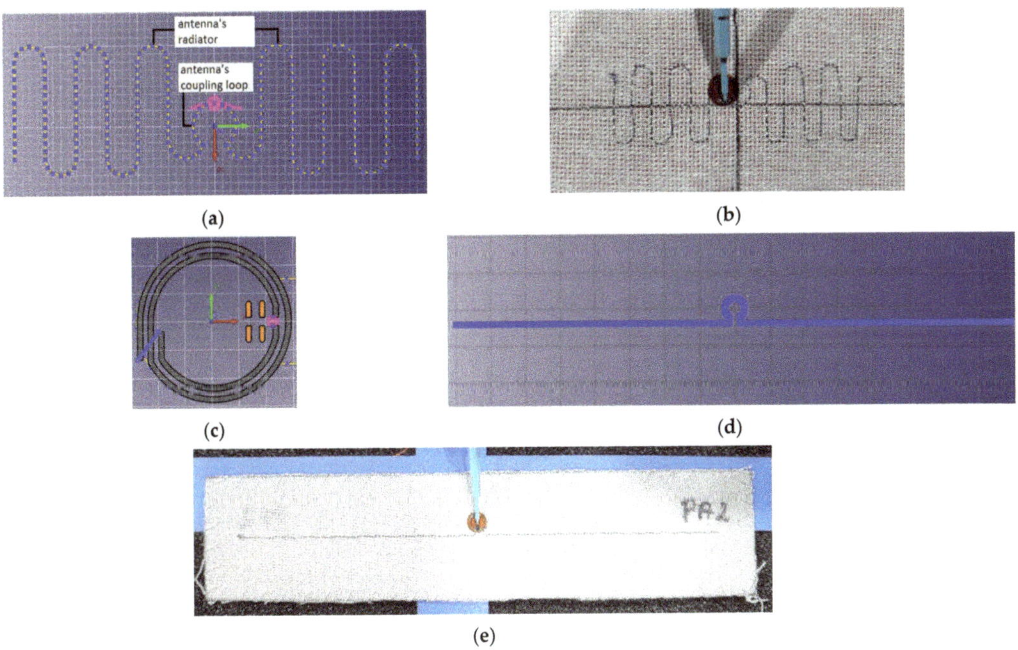

Figure 1. Textronic UHF RFID transponders: (**a**) EMCoS Studio simulation model of meander line antenna; (**b**) real sample of RFIDtex tag with meander antenna; (**c**) simulation model of chip coupling circuit; (**d**) EMCoS Studio simulation model of dipole antenna; (**e**) real sample of RFIDtex tag with dipole antenna.

The aim of the conducted research is to present factors impacting the performance of textronic UHF transponders. The main factors influencing the antenna impedance and chip voltage were determined. Among them were mutual inductance between the antenna and the chip coupling circuits and the thread resistivity. The simulation was conducted to establish its impact on the chip voltage and antenna impedance. Then, experimental research on the impact of the distance between coupling circuits, affecting mutual inductance, and embroidered antenna resistance on transponder antenna impedance and read range was performed.

The major issue in the further development of this concept is the lack of knowledge regarding crucial factors affecting the effectiveness and performance of the textronic UHF RFID transponder, their importance, and the relationships between them. The aim of this study is to identify these factors and to determine their impact on the transponder's operational parameters. The only research that has addressed this topic involves experimental studies on the influence of textile substrates [56] and the washing process [57]. Our investigation is the first attempt at a mathematical description of the textronic UHF RFID transponder's performance. The results obtained will allow us to define rules for designing and manufacturing properly working transponders according to specific application areas and required operational parameters. Likewise, they will shed light on development opportunities.

Development directions include both improving effectiveness and designing new transponder constructions that meet the condition of inductive coupling between the antenna and chip. The existence of various constructions of the textronic UHF RFID transponders is necessary in view of the nature of textiles and the clothing industry. The antenna geometry and microelectronic module shape are determined not only by their required performance but also by the available space on the fabric, its composition and type, the product designation, and aesthetics. The constructions considered in this paper consist of microelectronic modules manufactured as round elements like buttons or sequins attached plainly to fabric. However, such components are not used in all kinds of clothes.

The next interesting direction is the estimation of the RFIDtex tag's usefulness in automatic identification systems beyond the production and sales stages, including in Internet of Things implementations or as wearable RFID sensors. These application areas involve understanding transponder performance and how operational parameters change due to product use, integration, and destruction, as well as the impact of environmental conditions and the human body. It is presumed that the development and dissemination of this kind of wearable transponder could help advance the expansion of RFID systems in the textile industry and IoT, exploited at each step of production, sale, use, waste disposal, and recycling of textile products.

2. Materials and Methods

Firstly, mathematical calculations based on common electromagnetic laws were made in order to determine the relationships between the basic parameters of the transponder. The mathematical model was used to derive the formulas for antenna impedance, chip voltage, and mutual inductance between the coils of the coupling system. These expressions allowed us to obtain theoretical knowledge on the changes in electrical parameters depending on various factors. The conducted investigations led to the identification of research directions and the verification of laboratory measurements.

MATLAB was used to plot graphs illustrating the relationships between the considered factors based on the obtained relationships. In previous work on concept development, simulations were carried out using the EMCoS Studio 2022 software [55–57]. These outcomes were supplemented with measurements of magnetic field parameters. The next stage of the investigation involved experimental studies on the influence of two performance factors detected in the previous stage: the distance between the centers of the coupling circuits (both vertical and horizontal) and the embroidered antenna resistance.

The samples were sewn using an embroidery machine, BROTHER INNOV-IS V3. The maximum rotational speed of the embroidery machine's bobbin was 350 rpm. The tension of the upper thread (non-conductive), supporting the lower conductive thread, varied depending on the thickness and stiffness of the conductive thread. The bobbin also had an adjustable screw to regulate the tension of the thread, but this feature was not used during embroidery. The antenna impedance measurements were conducted using a Keysight PNA-X N5242A vector network analyzer, PacketMicro DPSS201505 SS05–0053 probe, and microscope to verify proper contact between the probe and the circuit. The read range measurements were performed in a Microwave Vision Group anechoic chamber equipped with a Voyantic Tagformance Pro measurement system with Tagformance UHF v.13.2.3 software (Figure 2).

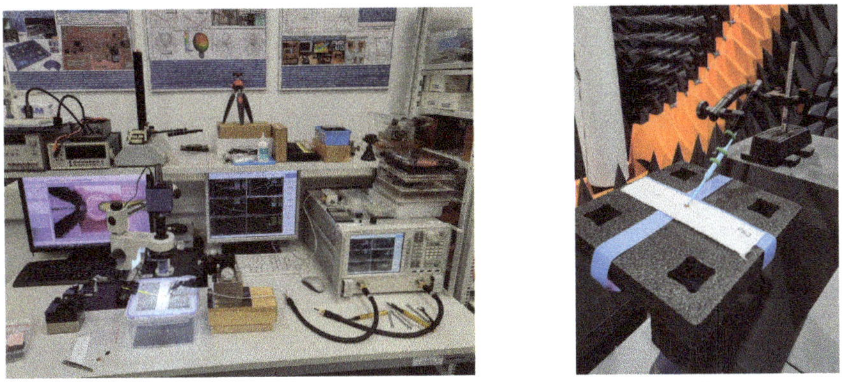

(a) (b)

Figure 2. Laboratory research stand: (**a**) stand for impedance measurements; (**b**) stand in anechoic chamber.

A epsilometer system with a Compass Technology measuring device and Vector Network Analyzer (VNA) Copper Mountain R60 was used to obtain the dielectric permittivity of the textile substrates.

3. Results

3.1. Calculations and Simulations

3.1.1. Mathematical Model of Textronic UHF RFID Transponder

The first stage of the investigation involved a mathematical description of the textronic UHF RFID transponder based on common electrical and electromagnetic laws. The research purpose was to determine the relationship between the complex impedance of the transponder components and the chip voltage. Both of these parameters refer directly to the operational parameters, such as impedance matching and the read range. The electrical equivalent of the radiofrequency frontend in the RFIDtex tag is presented in Figure 3.

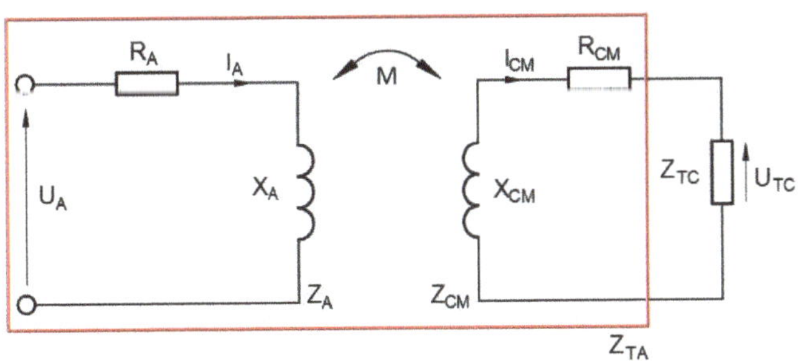

Figure 3. Electrical equivalent of the radiofrequency circuit in the textronic UHF RFID transponder.

The left part of the circuit diagram corresponds to the antenna module comprising the embroidered antenna and the coupling coil, where the impedance Z_A consists of the resistance R_A and reactance X_A:

$$Z_A = R_A + jX_A \tag{1}$$

The U_A value represents the voltage induced in the antenna module, and I_A is its current.

The second component of the transponder, the microelectronic module, forms the right circuit. It is combined with the chip, which has impedance Z_{TC}, and the coupling coil, whose impedance Z_{CM} is expressed as:

$$Z_{CM} = R_{CM} + jX_{CM} \tag{2}$$

The inductive coupling between both modules is described by mutual inductance M. Although the antenna module is sewn on fabric, the proper transponder antenna also includes a part of the microelectronic module. It is a combination of the left circuit described by Z_A and the chip coupling circuit (Z_{CM}) joined inductively, as indicated by the red line in the diagram. The impedance value of these components is required to determine the impedance matching between the chip and the transponder antenna.

Since the electric circuit diagram of the RFIDtex transponder (Figure 3) is similar to that of an air-core transformer, an equivalent circuit of a two-port network was created to obtain the desired relationships. The impedance Z_M in the equivalent circuit was expressed as an impedance consisting solely of the imaginary part X_M:

$$Z_M = jX_M \tag{3}$$

The following relationship for the antenna impedance was determined:

$$Z_{TA} = Z_{CM} + \frac{X_M^2}{Z_A} \tag{4}$$

Using the same equivalent circuit, the chip voltage U_{TC} can be obtained:

$$U_{TC} = I_{CM} \cdot Z_{TC} = \frac{U_A \cdot jX_M \cdot Z_{TC}}{Z_A(Z_{TC} + Z_{CM}) + X_M^2} \tag{5}$$

where I_{CM} is the current generated in the chip coupling circuit. The chip voltage is crucial in determining the power delivered to the chip. It influences the read range in the RFID system. For this reason, it is desirable to design RFIDtex transponders in a way that allows the generation of higher current in the microelectronic module than in the others.

In the procedure of obtaining the impedance Z_{TA} or the voltage U_{TC}, the main problem is to determine the value of X_M. In the case of the antenna impedance, the dependency on X_M is quadratic, but in the case of the chip voltage, the function exhibits a maximum. Attempts to obtain the optimal value of mutual inductance between coupling circuits will be the focus of future research. In the present investigations, various approaches to determine the value of X_M were taken into consideration, as described in the next section.

3.1.2. Mutual Inductance between Coupling Circuits

The mathematical model of the textronic UHF RFID transponder is a significant tool for addressing measurement issues related to its parameters. In the laboratory, only the impedance of the transponder antenna and the impedance of the chip coupling circuit need to be measured. Although the impedance of the embroidered antenna can be obtained as a result of numerical model computations in EMCoS Studio, there is no software feature to directly compute the impedance of mutual inductance.

The first approach to determining the value of X_M is based on the developed mathematical model. Equation (4) is rearranged into the following form:

$$X_M = \sqrt{(Z_{TA} - Z_{CM})Z_A} \tag{6}$$

The numerical values of the impedances used for further calculations can be obtained from numerical computations or laboratory measurements. In the presented research, only numerical computations are used. The limitation of this method arises from the accuracy of measurements and software simulations. However, its main advantages will become apparent when compared to the next two approaches, which use analytical electromagnetic formulas. These formulas are useful solely for simple geometries of the coupling system and only under certain assumptions.

In the investigations, two constructions of the RFIDtex transponder were considered: the first one with a 16 cm long dipole antenna and the second one with a meander line antenna. Both constructions have the same antenna and chip coupling circuits. The antenna coupling coil is a loop with a radius of 2.85 mm sewn in the middle of the radiator. The chip coupling coil consists of two connected concentric loops, with the external loop having a radius of 2.85 mm and the internal loop having a radius of 2.45 mm. In this basic case, the mutual inductance can be obtained through calculations based on common analytical formulas of electromagnetic laws. The first step is obtaining the magnetic flux density B according to the Biot–Savart law for a circular loop of radius r carrying a current i, expressed as [58]:

$$B = \frac{\mu_0 i}{2r} \tag{7}$$

It is assumed that the magnetic flux density is the same across the entire area of the loop and equals the value occurring at the center. That formula does not include the impact of the magnetic field generated by the arms of the antenna. Next, the magnetic flux is determined for the circular area of the second circuit as follows [59]:

$$\Phi_{12} = \iint_S B dS = \frac{\mu_0 i}{2} \pi r \tag{8}$$

And finally, the mutual inductance M is obtained [59]:

$$\mathrm{M}_{12} = \frac{\Phi_{12}}{i} = \frac{\mu_0}{2} \pi r \tag{9}$$

To calculate X_M, it was assumed that reactance is proportional to the frequency:

$$X_M = j2\pi f M \tag{10}$$

The third method was implemented to avoid some of these assumptions. It uses EMCoS Studio features to obtain some parameters of the magnetic field in the RFIDtex transponder components. The simulation was performed with U_A = 1 V. The magnetic flux density for the 16 cm long dipole antenna is presented in Figure 4.

The magnetic flux density of the chip coupling coil is shown in Figure 5.

From the graphs, it can be seen that B is not constant throughout the area, and for the chip coupling circuit, its value is highly diversified between the center and the edges. The use of a near field probe allows us to obtain the magnetic flux density at the selected points. The points are located on the diameter of the loop (Figure 6a), and the $\bar{B}$ value is obtained for each of them (Figure 6b).

The mutual reactance X_M obtained on the basis of the developed methods is presented in Figure 7.

As can be presumed based on the derived assumptions, the X_M value derived from the analytical formulas, assuming a smaller B value at the center, is the lowest one. Similarly, the approach with a higher B value at the edge gives excessive results. The values determined by the mathematical impedance model are located between them. From this comparison, the following conclusion may be drawn: using a mathematical model is a proper and reliable approach to obtain the X_M value. Other methods could also be used to approximate X_M, but only for simple circuit geometry.

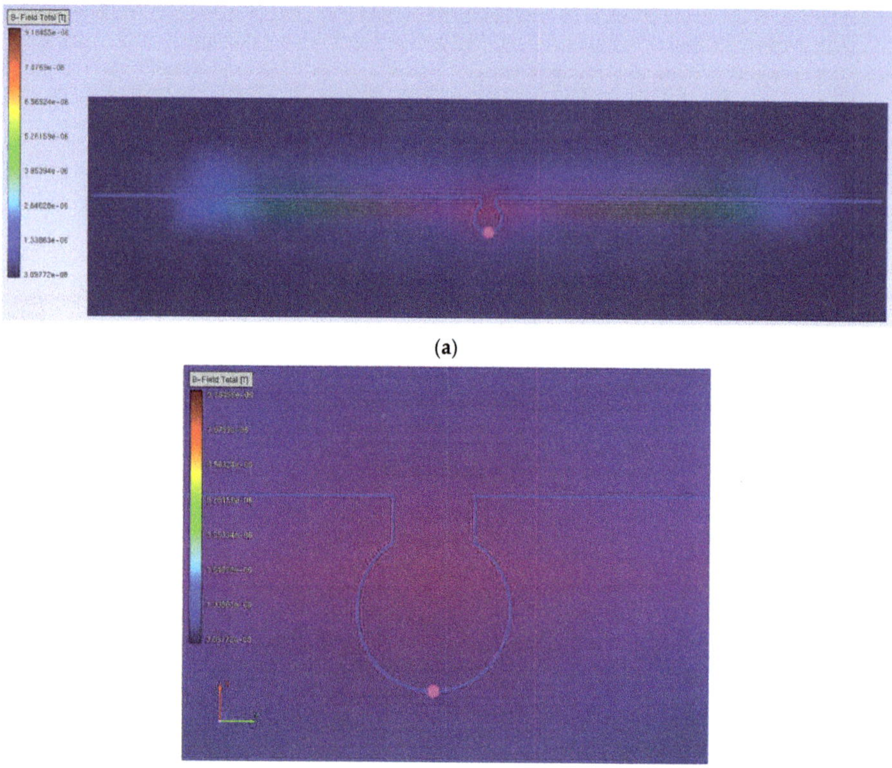

(**a**)

(**b**)

Figure 4. Magnetic flux density of 16 cm long dipole antenna for 860 MHz: (**a**) coupling circuit and radiator; (**b**) coupling circuit.

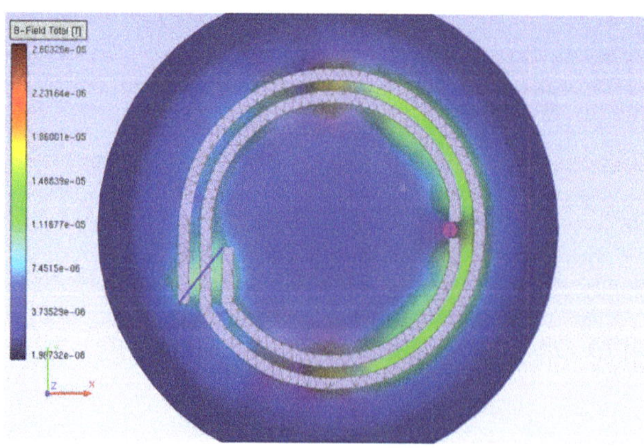

Figure 5. Magnetic flux density of chip coupling circuit for 860 MHz.

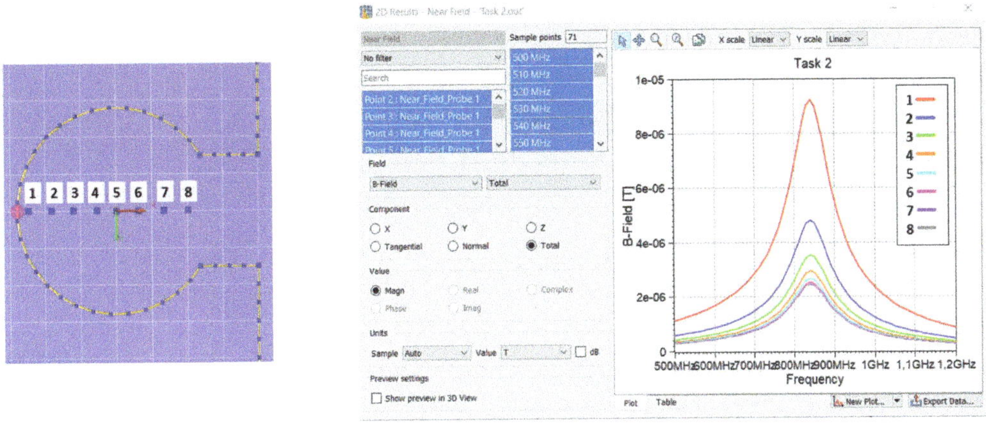

(a) (b)

Figure 6. Computation of magnetic flux density inside loop of coupling circuit: (**a**) location of measurement points; (**b**) results.

The highest value occurs at the point closest to the edge. In the third approach, this highest B value is used instead of analytical formula (7) to obtain the magnetic flux density. In this way, the drawback of previous approach is eliminated, where the calculated value of B is assumed in the center of the loop. As can be seen in Figure 7, this value is significantly lower at this point. Equation (8) is also used, but in (9), the embroidered antenna current I_A value is required. In the investigation, it is calculated from mathematical model relationships:

$$I_A = \frac{U_A(Z_{TC} + Z_{CM})}{Z_A(Z_{TC} + Z_{CM}) + X_M^2} \tag{11}$$

but it could also be computed by EMCoS Studio software.

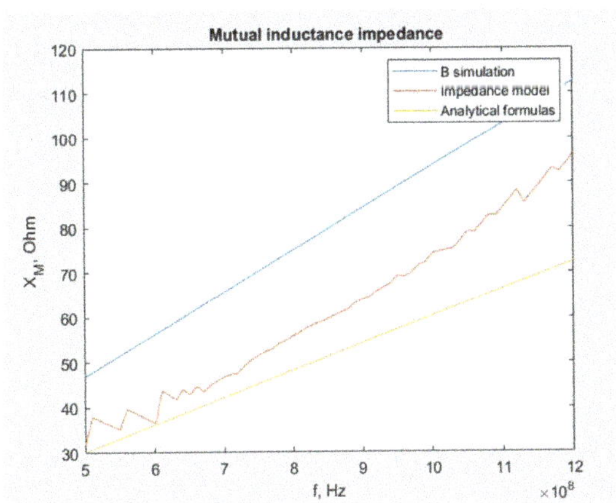

Figure 7. Comparison of different approaches used to obtain the X_M value for a 16 cm long dipole antenna.

The second design with the meander line antenna is an example of a more complex geometry, where methods based on the analytical description of electromagnetic parameters fail due to their assumptions. Although the coupling circuit is the same as before, its area is interfered with by the magnetic field derived from the antenna arms (Figure 8).

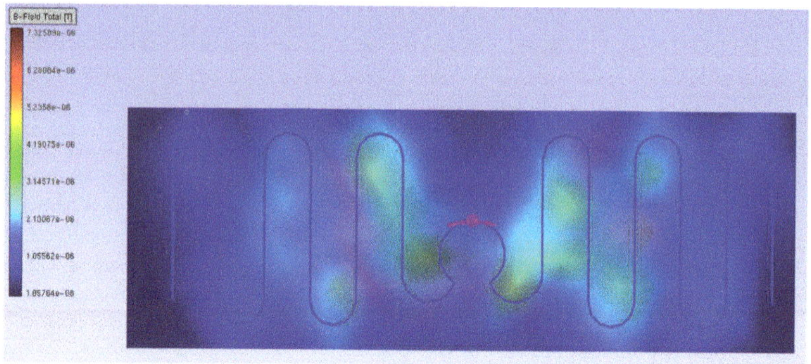

Figure 8. Magnetic flux density of the meander line antenna.

The same steps for the dipole antenna may be followed. The influence of the magnetic field generated by the antenna arms is noticed solely using the mathematical model (Figure 9).

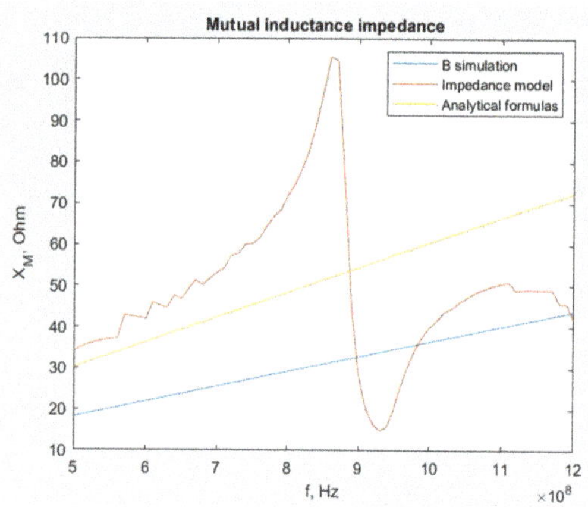

Figure 9. Comparison of different approaches used to obtain the X_M value for the meander line antenna.

For these reasons, in the next steps of this research, the assumed X_M value for each construction is that obtained by the mathematical model.

Another measure of magnetic coupling is the coupling coefficient, which qualitatively describes the coupling between circuits [60]. When the loops are centered on the same axis z, it can be determined as:

$$k(h) = \frac{r_1^2 \cdot r_2^2}{\sqrt{r_1 \cdot r_2} \cdot \sqrt{h^2 + r_1^2}^3} \tag{12}$$

where h is the distance between loops, as illustrated in Figure 10.

The coupling coefficient takes values from 0 to 1, where 0 means no coupling and 1 represents total coupling, assuming that the same magnetic field is passing through an area of both loops.

Equation (12) can be applied to approximate the impact of the vertical distance between the coupling circuits. This relationship is crucial for the RFIDtex transponder, especially due to possible failures in its manufacturing or the displacement of the fabric from the microelectronic module during garment use. For example, an elastic fabric with embroidery hung on a hanger may no longer adhere to the rigid button with the chip and its coupling circuit. As seen in Figure 11a, the coupling coefficient decreases dramatically at a distance of 2 mm; therefore, the RFIDtex transponder cannot work in the case of the above example.

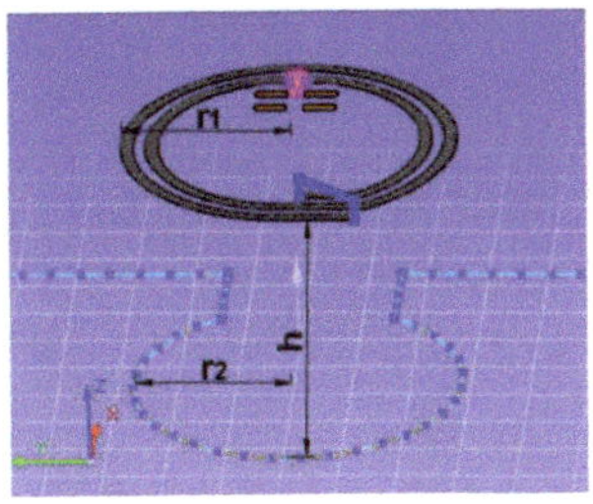

Figure 10. Geometry of coupling circuits.

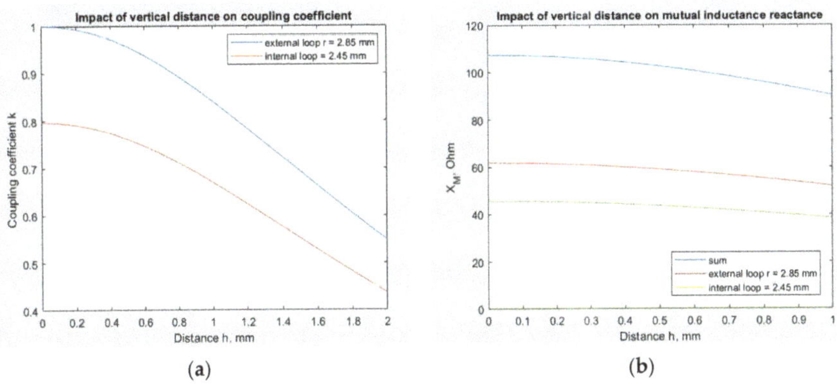

Figure 11. Impact of vertical distance on: (**a**) the coupling coefficient, (**b**) the reactance X_M of the RFIDtex transponder with a meander line antenna.

The influence of the coupling coefficient on the X_M reactance for a shorter distance of 1 mm in the transponder with the meander line antenna is shown in Figure 11b. The value of X_M for no distance is obtained from the mathematical model for a frequency of 866 MHz and equals 107 Ω. With increasing distance, the decrease in the X_M value becomes greater.

The mathematical model, combined with EMCoS Studio computations, allows us to simulate the impact of changing the X_M reactance on the chip voltage U_{TC}. The results obtained on the basis of Equation (5) for the dipole and the meander line antenna are presented in Figure 12. The impedance values are obtained from numerical computations, the chip impedance Z_{TC} equals 15.3—j313 (on the basis of Ucode 7m SL3S1214 datasheets by NXP Semiconductors), U_A is assumed to be 1 V, and X_M varies from 0 to 150 Ω based on the values obtained in the previous stage.

This large frequency bandwidth is displayed in the presented graphs because, at this stage of research, we were interested in observing the behavior of the U_{TC} function with changes in the selected variables, as well as identifying the position of its local and global maxima and minima. This knowledge is essential for investigating constructions of this type with inductively connected modules and for understanding the phenomena occurring within them. Only after completing this step it is possible to analyze the posed issues in the context of practical application in RFID technology. The ultimate goal of this study is to find the maximum U_{TC}. As can be observed, the voltage varies significantly with frequency, and for certain values of X_M, the maximum U_{TC} values are not achieved in the frequency band specific for RFID systems. Furthermore, at this stage of the research, a method for determining the geometric dimensions of the transponder elements to achieve resonance has not yet been developed. For this reason, it is beneficial in analyzing the results of transponder simulations and measurements, both current and future, to have calculations for a broader frequency range. Of course, in the future, the investigated frequency range will be gradually limited until only standard frequencies become the subject of interest.

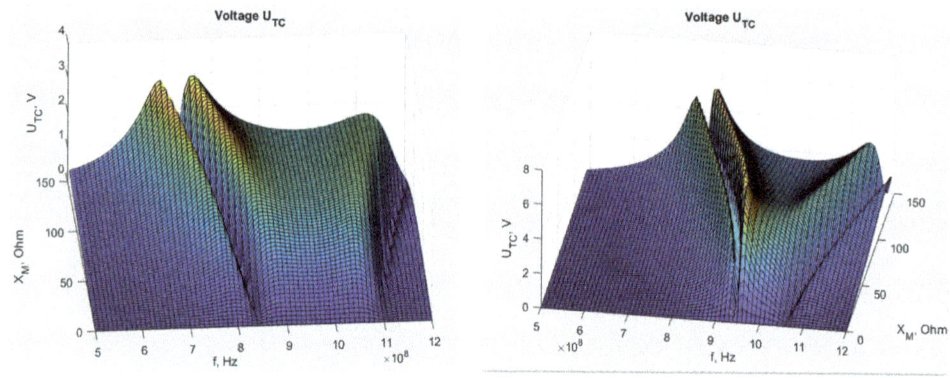

(**a**) (**b**)

Figure 12. Chip voltage dependency on reactance X_M: (**a**) transponder with dipole antenna; (**b**) transponder with meander line antenna.

The different shapes of the graphs show that the antenna geometry has a significant impact on the chip voltage, and consequently, on the read range. For the meander line antenna, the maximum voltage is higher than for the dipole, but in that case, these maximal values dramatically decrease with variations in the X_M reactance. This stronger dependency can indicate that the meander line antenna is less stable in the case of unintentional changes in its construction, such as displacement of the microelectronic module relative to the antenna coupling circuit, and its performance may significantly degenerate. This implies better performance but also greater vulnerability to manufacturing inaccuracies or usage destructions. However, these conclusions are valid only for certain values of X_M, because, as seen in Figure 12, there are regions of milder and stronger changes. Not necessarily for every value of X_M, the region of strong changes for the meandered antenna corresponds to the region of milder changes for the dipole antenna.

From the viewpoint of further development of these constructions, it can be stated that the tag with a dipole antenna and $X_M = 63\ \Omega$ at frequency 866 MHz is far from its optimal point, and there is a need to improve the efficiency of its coupling circuit. In contrast, the meander line antenna with $X_M = 107\ \Omega$ at frequency 866 MHz is close to the optimal point, and its X_M value is excessive. In this case, the deterioration of the coupling circuit efficiency may cause performance improvement. However, considering the assumptions made regarding calculations and simulations, the obtained numerical values are approximated and may lead to incorrect conclusions about the search for the optimal value for a given construction. This issue will be investigated in the future.

The next factor affecting the RFIDtex transponder is the resistivity of threads used in the antenna embroidering. The real part of the embroidered antenna impedance can be calculated based on its length and thread resistivity. The relationship between the chip voltage and the antenna resistance can be obtained using Equation (5) and the same numerical computations (Figure 13).

The graph shape also depends on the antenna geometry. The significant influence of resistance is observed for the dipole antenna in the range of 800–900 MHz, where the chip voltage decreases with increasing R_A. However, at the same time, the decrease itself becomes smaller until a certain value of R_A, and then the value stabilizes. For the meander line antenna, the same phenomenon occurs for the band around 970 MHz. For other frequencies, R_A has no great impact on U_{TC}.

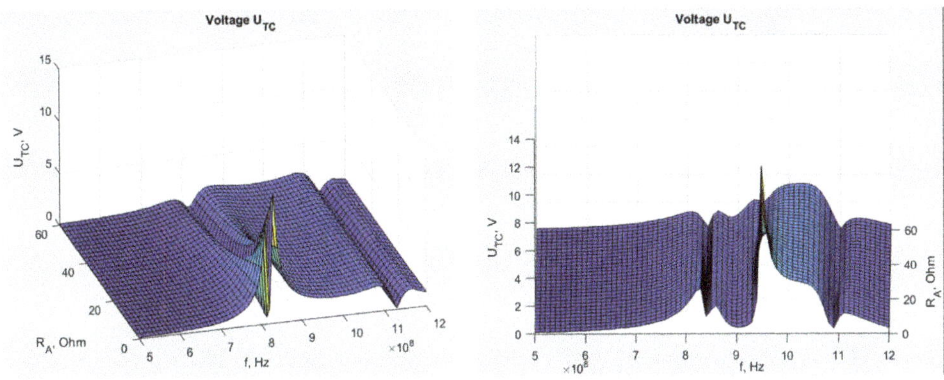

(a) (b)

Figure 13. Chip voltage dependency on embroidered antenna resistance: (**a**) transponder with dipole antenna; (**b**) transponder with meander line antenna.

Equation (4) is used to obtain the dependency of transponder antenna impedance Z_{TA} on the embroidered antenna resistance R_A (Figures 14 and 15).

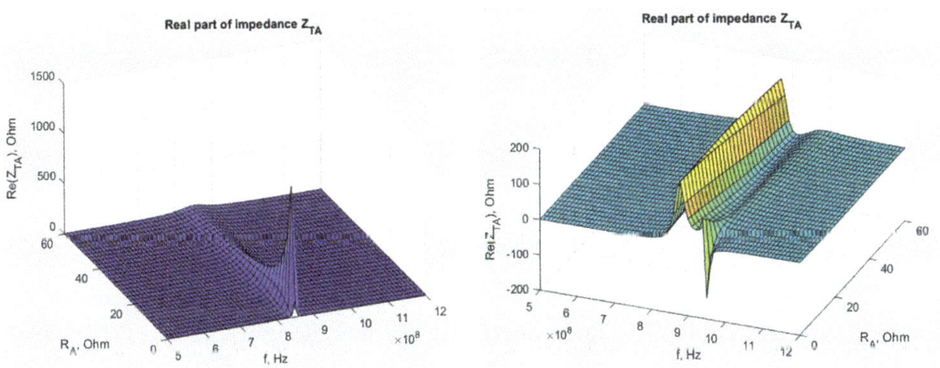

(a) (b)

Figure 14. Real part of transponder antenna impedance dependency on embroidered antenna resistance: (**a**) transponder with dipole antenna; (**b**) transponder with meander line antenna.

The influence of the antenna geometry is also crucial. The real part of transponder impedance decreases for the dipole antenna, and later, it is constant at increasing R_A values. However, this dependency is more complex for the meander line antenna, and a small or no impact is also observed at most frequency values.

3.2. Experimental Research

3.2.1. Impact of Vertical Distance between Antenna and Chip Coupling Circuits

For the experimental investigation, the meander line antenna analyzed in Section 3.1 was used. It was sewn using Agsis Syscom thread with a global thread tension of 0 and a tension of 9. The distance between the coils in the coupling system was increasing by inserting pieces of Kapton between them. Each piece was 125 μm thick.

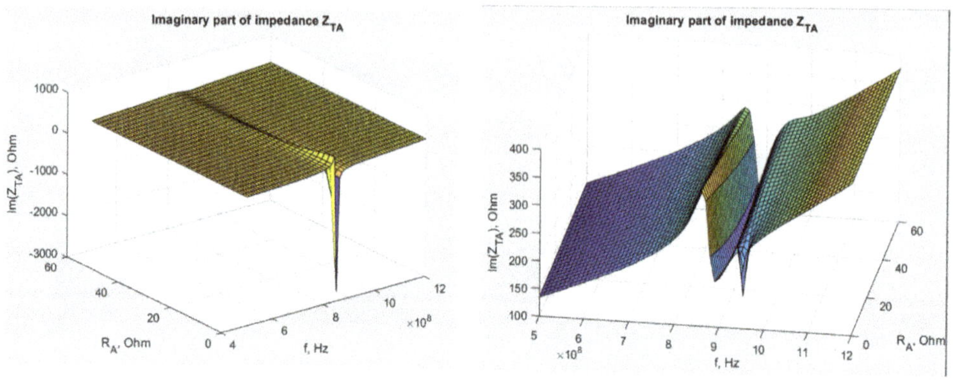

(a) (b)

Figure 15. Imaginary part of transponder antenna impedance dependency on embroidered antenna resistance: (**a**) transponder with dipole antenna; (**b**) transponder meander line antenna.

The first step of the investigation involved measuring the dielectric permittivity of Kapton using the epsilometer (Figure 16).

(a) (b)

Figure 16. Laboratory research stand: (**a**) epsilometer; (**b**) piece of Kapton under testing conditions.

The measurement was repeated for each subsequent piece of Kapton placed on the remaining stack. This way, spacers with thicknesses of 125, 250, 500, 750, and 1000 μm were achieved. The relative dielectric permittivity for each separator is presented in Figure 17.

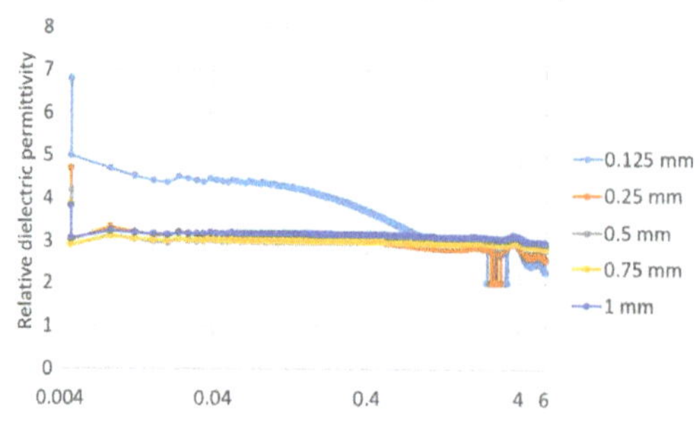

Figure 17. Relative dielectric permittivity of Kapton spacers.

Except for the first case for the single Kapton layer, the relative dielectric permittivity was similar for subsequently increasing spacers. The tangent of the loss angle for each separator is shown in Figure 18. As the thickness increases, smaller changes in the tangent of the loss angle are observed.

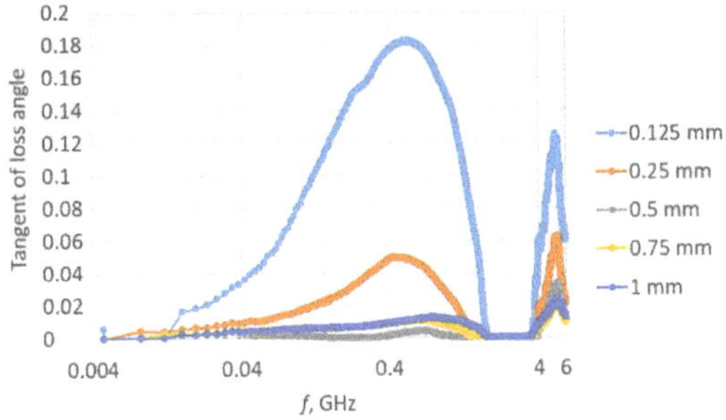

Figure 18. Tangent of the loss angle of Kapton separators.

In the next experiment, the transponder antenna impedance measurements were conducted for each separator and with no distance between the coupling circuits (without any Kapton layer). The test setups are shown in Figure 19, and the results are shown in Figure 20.

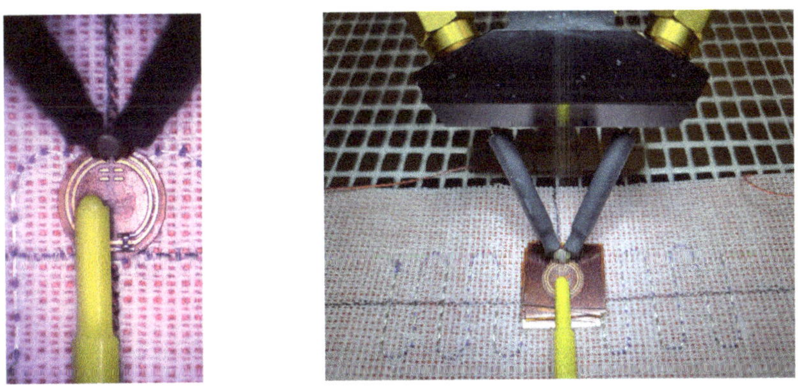

(**a**) (**b**)

Figure 19. Measurements of vertical distance impact: (**a**) coupling circuits exactly aligned with no vertical or horizontal distance; (**b**) coupling circuits with a vertical separator.

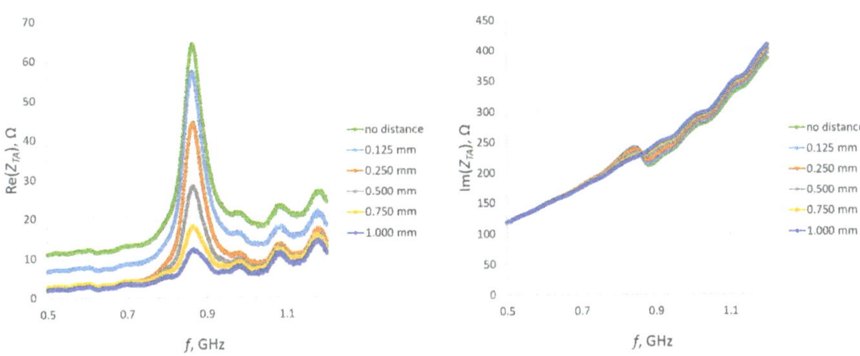

(**a**) (**b**)

Figure 20. Impedance of transponder antennas Z_{TA}: (**a**) real part R_{TA}; (**b**) imaginary part X_{TA}.

The relationship between the vertical distance and the real part of the impedance is consistent with the numerical calculations shown in Figure 11. The larger the distance between the coupling circuits, the lower the magnitude in the real part of the impedance. Also, the magnitude of decrease is lower. The impact on the imaginary part of impedance is also visible. With increasing distance, the resonance becomes smaller, until it eventually disappears.

The final stage involved measurements of the read range (Figure 21). The parameters of the Voyantic sweep settings were as follows: minimum power of −5 dBm, maximum power of 25 dBm with a step of 1 dB, start frequency of 800 MHz, and stop frequency of 1000 MHz with a step of 10 MHz. The sweep direction was rising. The transmitter had an output power of 29 dBm, and its antenna had a maximum gain of 6 dBi. The sensitivity of the receiver was −70 dBm, and its antenna had a 6 dBi gain. The ISO 18000-6C communication protocol was applied with the query command. The forward link was 25 µs DSB-ASK, and the return link was FM0, 40 kHz. The wideband UHF reference Tag v1 delivered with the Voyantic system was used to set up the measurements.

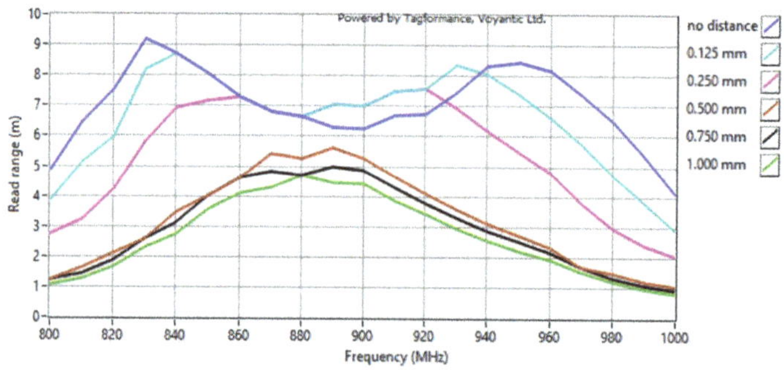

Figure 21. Read range of RFIDtex transponders with a vertical distance between the microelectronic module circuit and the antenna coupling circuit.

The shape of the waveforms obtained for the separator cases: no distance, 0.125 mm and 0.250 mm, is consistent with the surface of the graph for U_{TC} dependency in Figure 12b. The different shapes of the remaining curves may result from the attenuation of the magnetic field influence generated by the meander line antenna with increasing distance. In consequence, the obtained results are more similar to the surface of the respective graph for the dipole antenna (Figure 12a), and the decrease is not as dramatic as for the meander line antenna or the read range for smaller distances. The disappearance of the two extremes is also noticeable with the decrease in X_M related to the increase in the distance.

In light of the earlier considerations, the decrease in X_M should result in better transponder performance, because the optimal value is lower than achieved by this construction. However, this phenomenon is not observed. This could have happened due to the assumptions made in the calculations, causing the obtained values to be approximate, or due to the loss of the original geometry as the circuits moved away from each other. This issue must be further investigated in the future.

3.2.2. Impact of Horizontal Distance between Antenna and Chip Coupling Circuits

During manufacturing or use of clothes, horizontal displacements of the circuits can also occur. Such displacements cause variations in the coupling coefficient and mutual inductance as a result of changes in the common surface area of the coupling coils and limitations in the magnetic flux.

The same structure of the RFIDtex transponder is examined in this investigation. The four cases of displacement are considered relative to the center of the antenna radiator—shifts downwards, upwards, left, and right (Figure 22). The assumed distance between the centers of the coupling circuits is equal to the length of the loop radius (circa 2.85 mm).

For each case, the transponder antenna impedance is measured (Figure 23). Next, the read range for each displacement is obtained (Figure 24).

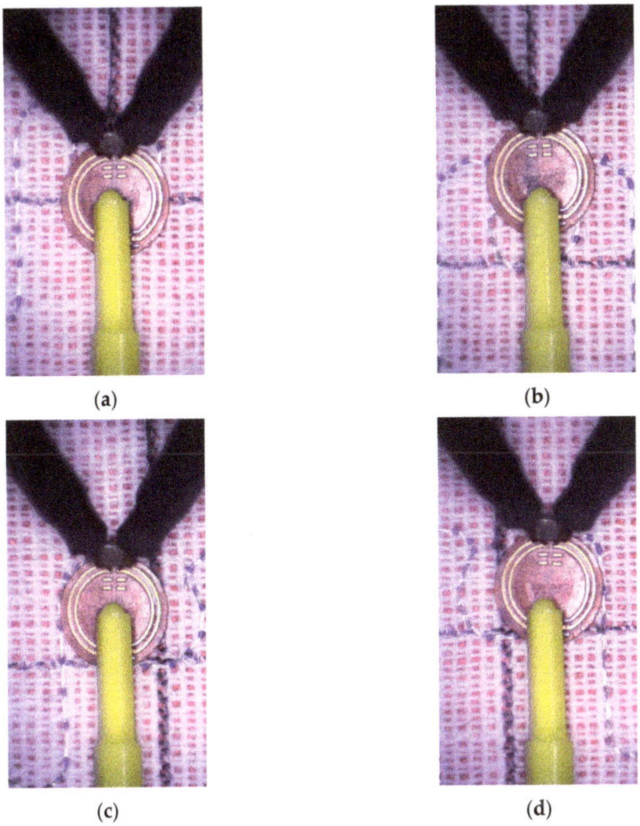

Figure 22. Location of chip coupling circuit relative to antenna coupling circuit: (**a**) moved down; (**b**) moved up; (**c**) moved left; (**d**) moved right.

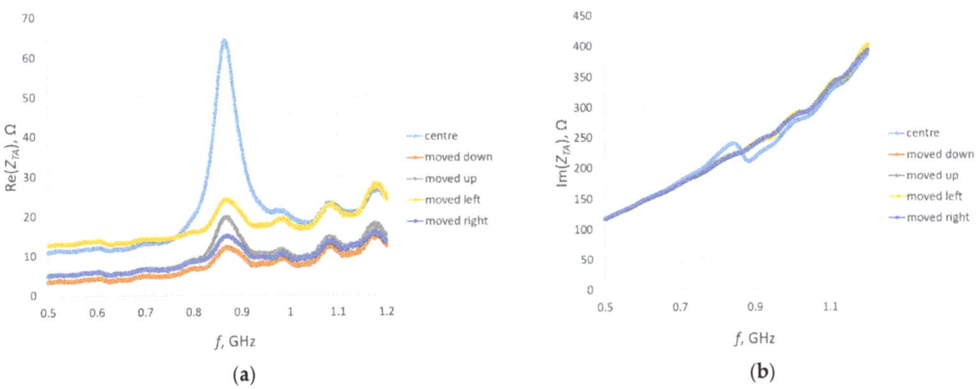

Figure 23. Impedance of transponder antennas Z_{TA}: (**a**) real part R_{TA}; (**b**) imaginary part X_{TA}.

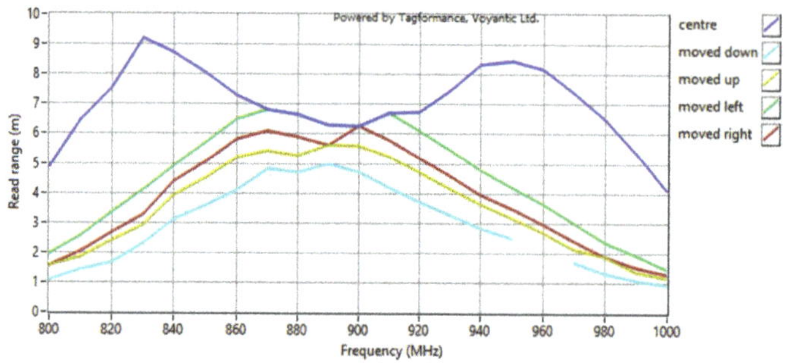

Figure 24. Read range of textronic UHF RFID transponders with different locations of the microelectronic module relative to the antenna coupling circuit.

The obtained results of the read range correspond to the values of the magnetic flux density presented in Figure 8. The meander line antenna is symmetrical, and in this regard, the right and left sides can be interchanged. When simulating the magnetic field distribution, the same results for shifts to the right and left might be expected due to the symmetry of the antenna. However, the direction of the shift also matters in both the computer calculations and the laboratory measurements, which are consistent with each other.

In summary, the results of the study indicate a greater vulnerability to vertical displacement when examining the impact of vertical and horizontal shifts. The same or even lower values were obtained for vertical changes compared to those observed for the experiments in the horizontal direction. However, the maximum vertical distance was 1 mm, whereas the maximum horizontal shift equaled approximately 2.85 mm. This is indeed a disadvantage of the proposed RFIDtex tag design, as distances as small as 1 mm may occur due to even small thread pulls in the antenna embroidery or fabric beneath the microelectronic module. The microelectronic module should also be precisely sewn to the fabric to ensure that the antenna fits tightly to its rigid construction. It can be also observed, based on the analysis of the relationship between the chip voltage U_{TC} and the reactance X_M, as well as the distribution of the magnetic field, that the transponder with the dipole antenna exhibits less sensitivity to coupling system displacements than the construction with the meander line antenna.

3.2.3. Impact of Antenna Resistance

The influence of the embroidered antenna resistance R_A is elaborated upon in the subsequent investigation. The R_A depends directly on the thread resistivity and the antenna length. The same 16 cm long dipole antenna is used as in the simulation in Section 3.1.2. The same design is embroidered with nine various conductive threads. The antenna resistance is calculated based on the thread resistivity provided by the producers (Table 1).

Table 1. Resistivity of conductive threads and resistance of embroidered 16 cm long dipole antenna.

Group Name	Thread	Thread Resistivity, Ω/m	Antenna Resistance R_A, Ω
PA	AGSIS SYSCOM	82.00	13.12
PB	ADAFRUIT 603	39.37	6.30
PC	SPARKFUN DEV-11791	91.86	14.70
PD	ADAFRUIT 641	32.68	5.23
PE	ADAFRUIT 640	51.18	8.19
PF	ELECTRO FASHION	40.00	6.40
PG	LIBERATOR 40	3.28	0.52
PH	INNTEX PW018A	300	48.00
PJ	LICA 10 × 0.04 mm	1.39	0.22

For each thread, the same antenna was sewn several times. The samples of the PA group were embroidered with a global thread tension of 0 and a tension of 9. For groups PC, PE, PF, PG, PH, and samples PD1-PD3 and PJ1-PJ3, PB1-PB3, the tension was set to 9 and the global tension was set to 8. For samples PD4, PJ4, PB4, and PB5, the global tension was changed to the value of 0.

For each sample, a microelectronic module without a chip was attached to measure the impedance Z_{TA} of the transponder antenna. The measurements are presented in Figures 25 and 26.

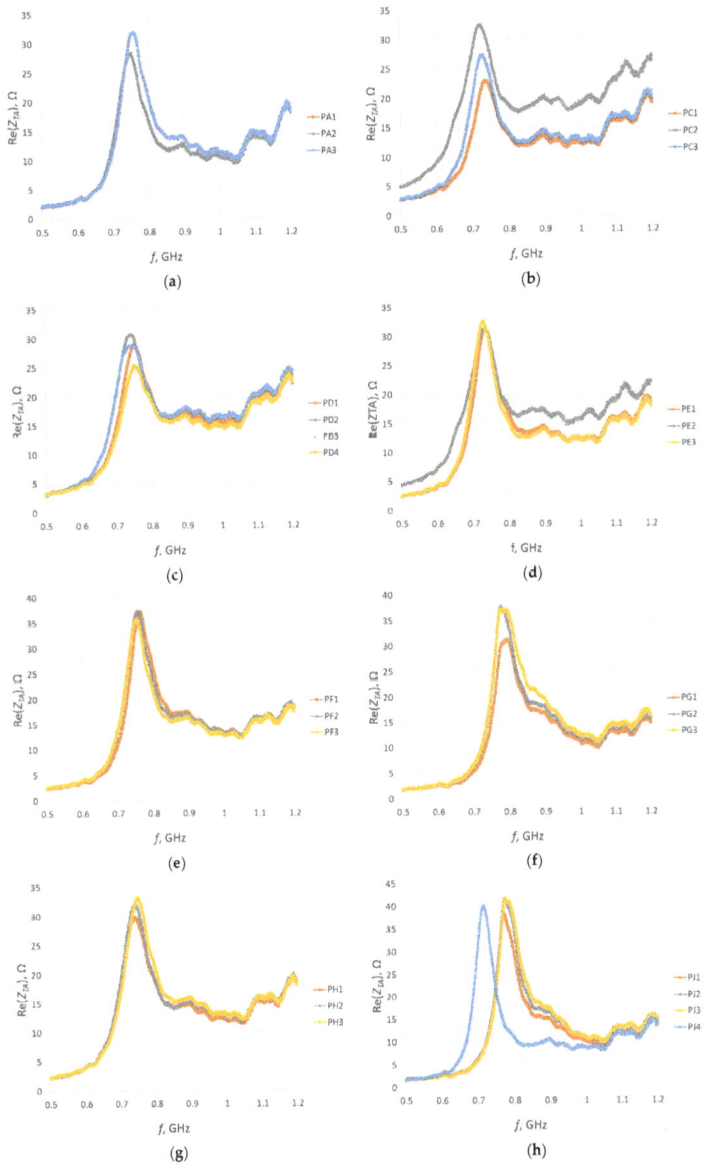

Figure 25. Real part of the transponder antenna impedance for the following groups: (**a**) PA; (**b**) PC; (**c**) PD; (**d**) PE; (**e**) PF; (**f**) PG; (**g**) PH; (**h**) PJ.

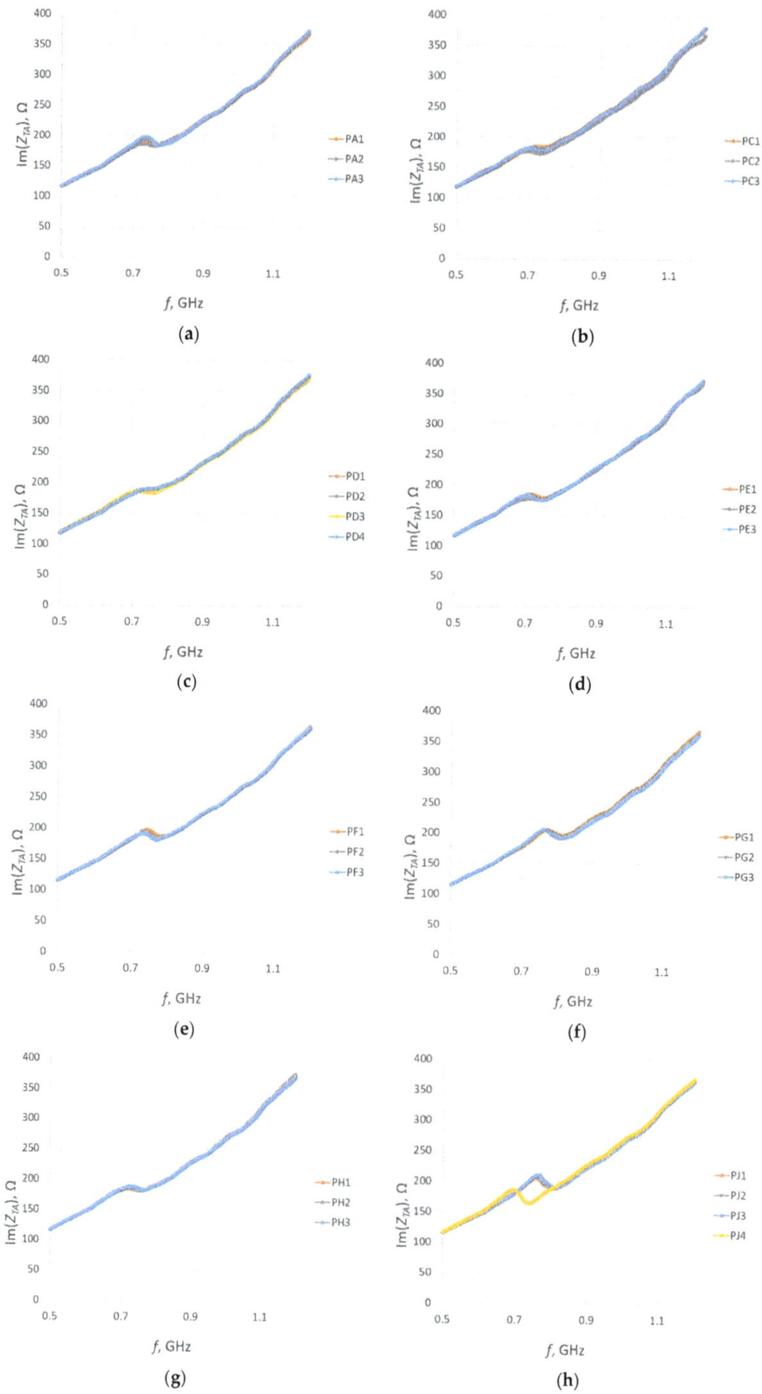

Figure 26. Imaginary part of the transponder antenna impedance for the following groups: (**a**) PA; (**b**) PC; (**c**) PD; (**d**) PE; (**e**) PF; (**f**) PG; (**g**) PH; (**h**) PJ.

In the next step, for each sample, the microelectronic module with the chip is appended, and the read range is measured (Figure 27). The parameters of the measurement process were the same as in the previous study, as described in Section 3.2.1.

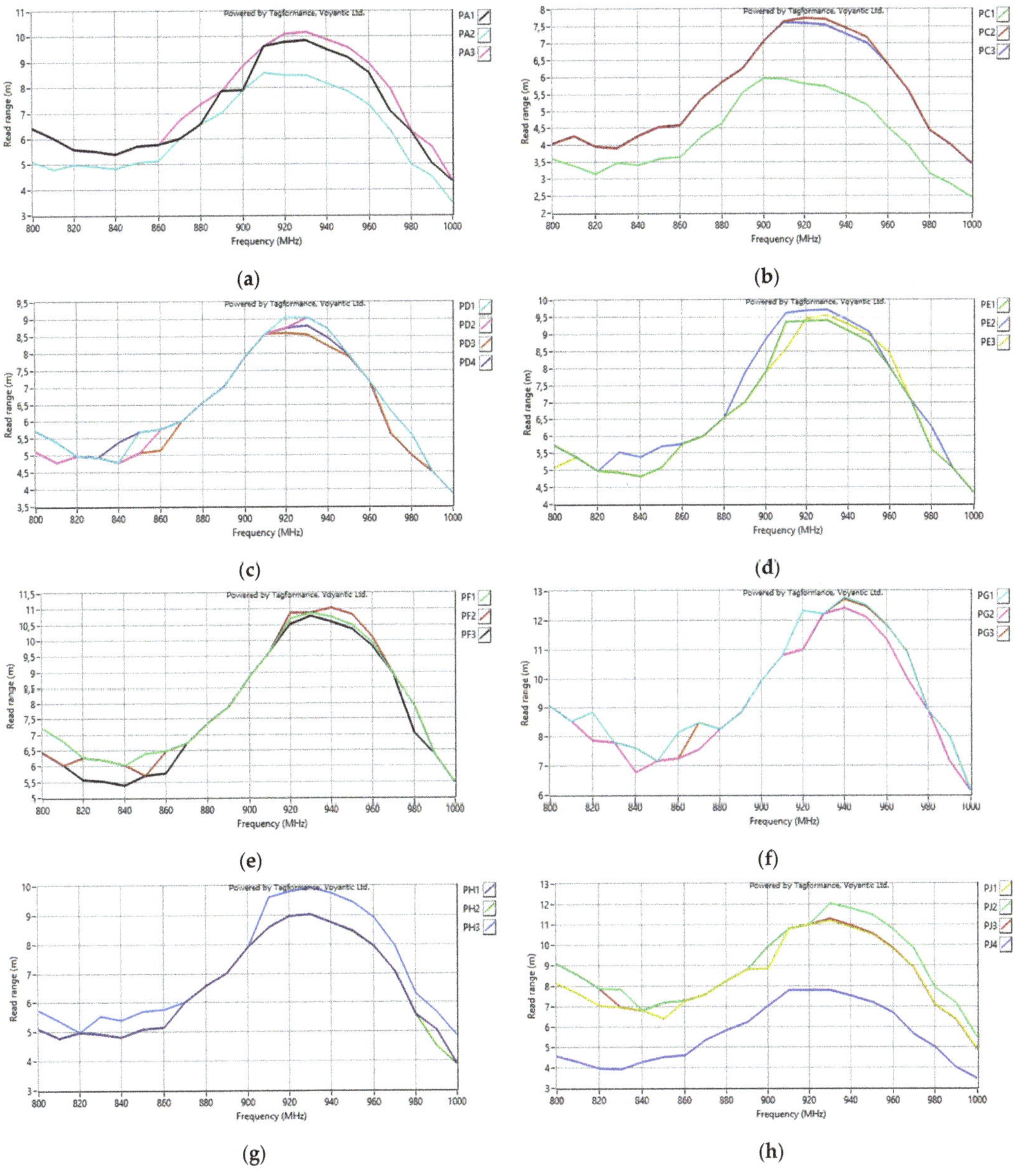

Figure 27. Read range of the transponder antennas for the following groups: (**a**) PA; (**b**) PC; (**c**) PD; (**d**) PE; (**e**) PF; (**f**) PG; (**g**) PH; (**h**) PJ.

The antennas from groups PB and PJ4 were omitted in the analysis of measurements due to failures in embroidering the geometry. The fabricating issues are also presumed to provoke shifts in the resonant frequency. This is described in detail in Section 3.2.4.

The decrease in the impedance in some groups is probably the result of inaccurate alignment in the coupling systems. This impact is presented in Section 3.2.2. The exact alignment is difficult to maintain under laboratory conditions because, during measurements, the microelectronic module is not permanently sewn to the antenna. The microelectronic module is placed on the antenna module, and the alignment with the coupling circuit is performed manually each time.

To compare different threads, one sample was chosen from each group (excluding PB) for which, within its respective group, the read range was the largest. A comparison of the impedance Z_{TA} of these samples is shown in Figure 28.

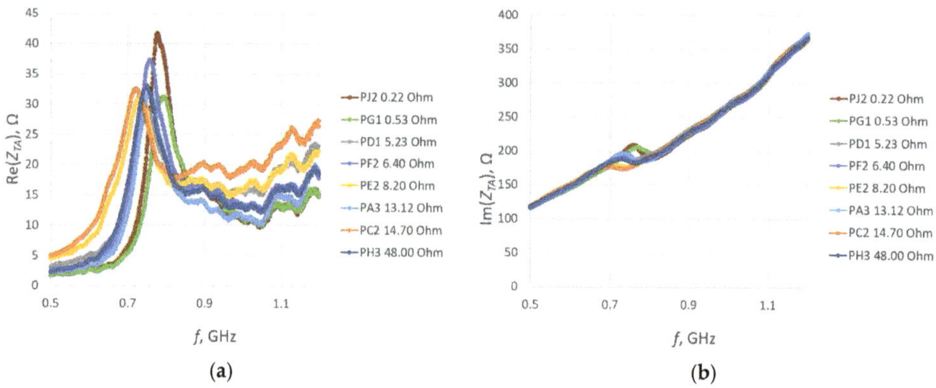

Figure 28. Impedance of transponder antennas sewn using different threads: (**a**) real part; (**b**) imaginary part.

The obtained results strongly correspond to the graph in Figure 14a. Shifts in the resistance were observed between groups, what is considered to be caused by the manufacturing issues analyzed in Section 3.2.4.

The comparison of the read range for the samples sewn using various threads is presented in Figure 29.

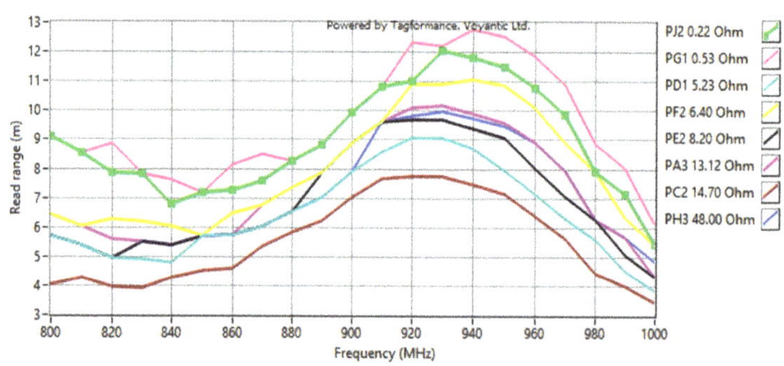

Figure 29. Read range of RFIDtex transponders sewn using different threads.

The accuracy of both the impedance and read range measurements may be influenced by the positioning of the microelectronic system. Conversely, the interpretation of the obtained results may be disrupted due to incorrect data on the thread resistivity provided by the producers. As can be seen from the graphs, the chip voltage of the PD sample was lower than expected when compared to the results for groups with the resistance R_A of the closest value. This may suggest that the resistivity of the thread used in the PD group can be higher than the value specified by the producer.

The read range values reached between the groups correspond with the graph shown in Figure 13a. Nevertheless, the maximum values are shifted towards higher frequencies.

The influence of the embroidered antenna resistance is limited to a certain value, which means that above this threshold, there is little to no change in the transponder performance.

3.2.4. Vulnerability to Manufacturing Issues

The samples from the PB group sewn using Adafruit 603 thread are omitted in the analysis of manufacturing issues. The reason for this decision is the discrepancy in the obtained measurements for this sample (Figure 30). This fault indicates poor antenna embroideries, which are caused by technical difficulties in the manufacturing process. Interestingly, no variation in the read range is observed with respect to frequency. This result is surprising, as the other parameters are unstable; however, it is consistent with the expected results based on the relationship between U_{TC} and R_A, described in Section 3.1.2.

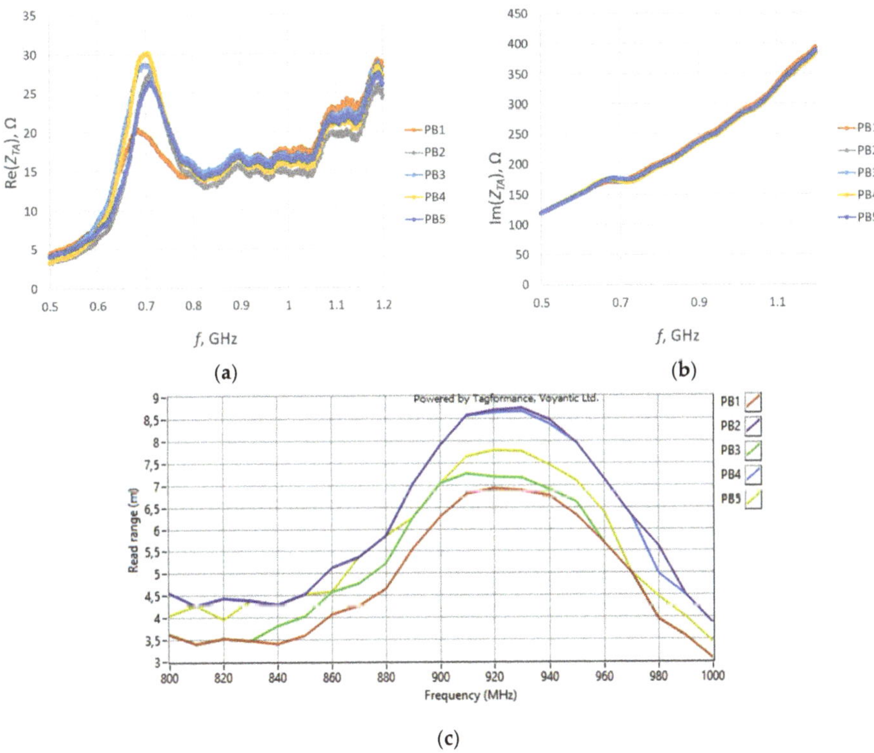

Figure 30. Measurements of group PB: (**a**) real part of antenna impedance; (**b**) imaginary part of antenna impedance; (**c**) read range of textronic UHF RFID transponders.

The reason for the differences between the charts lies in the faulty stitching of the antennas. In Figure 31, the samples for groups PA–PC are presented. The upper thread in the embroidery machine is a regular blue color. It supports the grey conductive thread, which serves as the lower thread in the machine. The visible blue thread knots at the ends of the radiators on the scans do not constitute part of the antenna's conducting element.

The grey conductive thread in the PB group significantly tangles, making it impossible to stitch the desired shape, both small loops and long straight segments. This is because the thread is too thick in relation to the parameters of the used embroidery machine. Although the maximum read range measurements obtained in this group are higher than the worst results from correctly sewn samples, the discrepancies between the samples in the PB group prevent the use of this thread. The precise and repeated stitching of the same geometry using this model of embroidery machine is not possible.

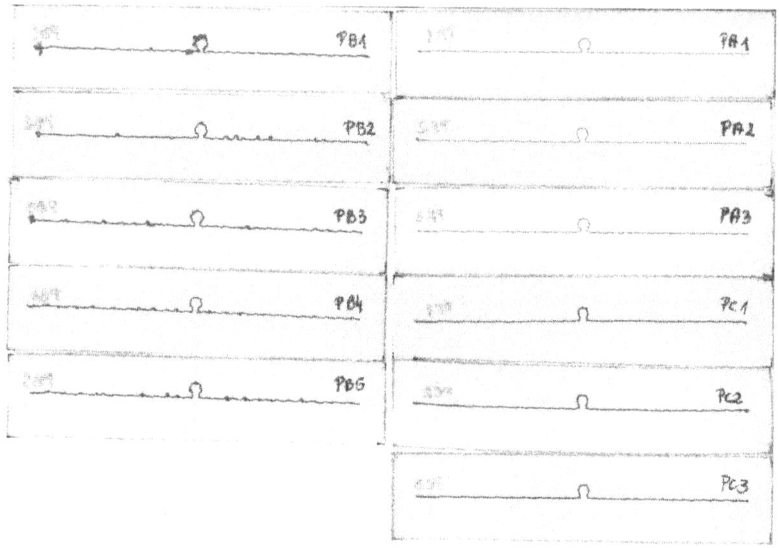

Figure 31. Scans of textronic UHF RFID transponders sewn using different threads; groups PA, PB, and PC.

As can be seen in Figure 25b or Figure 26b, the resonant frequency of sample PC2 is lower than that of the other antennas. The cause of this can also be attributed to the quality of the sample's fabrication. The right arm of the PC2 antenna is straight, but the left one is clearly curved. In Figure 32, a straight red line is marked, from which the left arm of the antenna deviates.

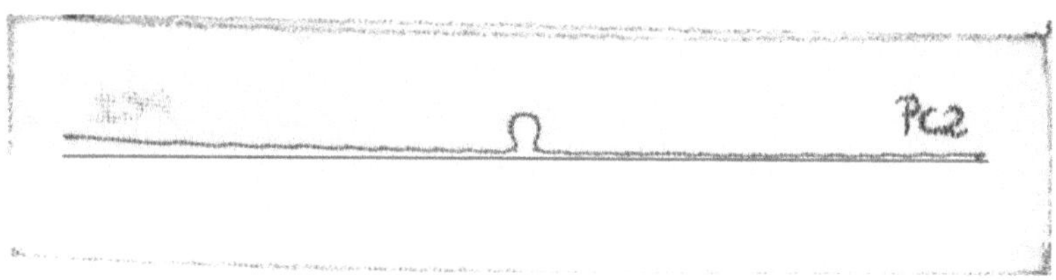

Figure 32. Curvature from the straight line of the left arm of the PC2 antenna.

The samples of the antennas from groups PD–PG are shown in Figure 33. It can be observed in Figure 25c that the PD2 and PD3 antennas have a lower resonant frequency than the others. As seen in the scans, the left arm of these antennas is noticeably wavier compared to samples PD1 and PD4.

The last two groups of RFIDtex tags are presented in Figure 34. As seen from the graphs in Figures 25h and 26h, the measurements for sample PJ4 significantly differ from the others.

In the PJ4 antenna, grey loops of conductive thread protruding from the embroidery can be seen. These result from an attempt to embroider the antenna using a conductive thread that has an excessively high stiffness.

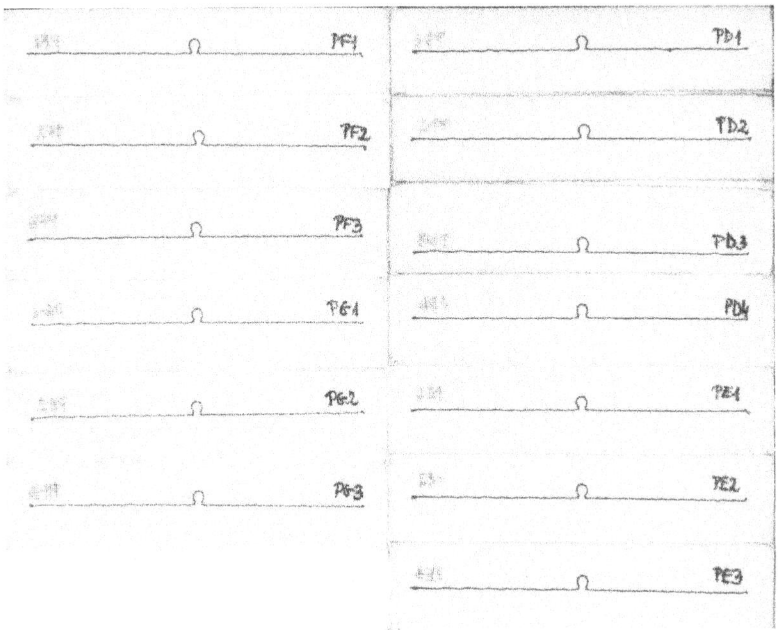

Figure 33. Scans of textronic UHF RFID transponders sewn using different threads; groups PD, PE, PF, and PG.

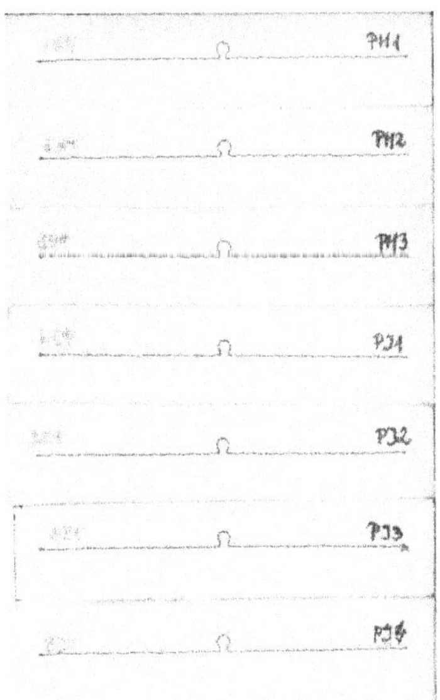

Figure 34. Scans of textronic UHF RFID transponders sewn using different threads; groups PH and PJ.

4. Discussion

The significant dependence on the distance between the centers of coupling coils is concerning, because even small deviations prompt substantial changes in the operational parameters. These deviations can be caused, for example, by the loosening of threads around the sewn button (microelectronic module), its displacement, threads under the button, or material wrinkling. Although the product is manufactured correctly, if the RFIDtex tag is intended to be used throughout the product's lifecycle, such damage is highly probable to occur over time. Based on numerical calculations, it is presumed that the high sensitivity to such displacements is inherent in construction with the meander line antenna. Applying a different geometry of the antenna module should overcome this problem, which will be investigated in the future.

The conducted research shed light onto an interesting feature of the textronic UHF RFID transponder, i.e., its dependency on the embroidered antenna resistance. A lower thread resistivity results in a higher chip voltage and transponder antenna impedance. However, there is a specific resistivity value for the thread beyond which its variations do not significantly impact the change in the impedance or the induced voltage at the chip terminals. In RFID systems where smaller ranges are sufficient, the resistivity ceases to be crucial in selecting an appropriate thread (or another material suitable to manufacture a textile antenna). This is especially useful, considering the challenges in measuring thread resistivity and the possibility of incomplete or inaccurate information provided by manufacturers. This feature also indicates that a deterioration in thread resistivity during product usage may not alter the transponder's operational parameters. And in the case of low resistivity threads, it might be possible to estimate such changes. However, it should be emphasized that this conclusion is correct under the assumption that product usage will not adversely affect other thread characteristics such as changes in the antenna geometry or mechanical injuries.

In light of the above considerations, when designing an RFIDtex transponder, a greater challenge than selecting a thread with an appropriate conductivity or monitoring its changes over time is ensuring accuracy in the fabrication of the antennas. Even minor thread pulls or stitch curvatures caused significant deviations from the expected results. Therefore, in systems where threads with lower conductivities provide sufficient transponder performance, the chosen thread should be selected for ease of use in embroidering the correct geometry and ensuring repeatability of measurements for subsequent samples.

The next step in the identification of factors affecting the performance of the textronic UHF RFID transponder was the exploration of various geometries of antenna radiators and coupling circuits. At this stage of the research, significant differences in the operation of transponders were observed solely due to the application of different antenna geometries in their structures. The strong influence of geometry on the transponder's efficiency is further evidenced by issues arising from incorrect stitching or imprecise placement of the microelectronic circuit in relation to the antenna. Thus, different dimensions and shapes of coupling systems should be taken into account. It will be necessary to develop methods that allow for the selection of coupling system dimensions so that, for given antenna and module shapes, it will be possible to approach the maximum of the chip voltage. For the developed microelectronic module designs, corresponding forms of clothing elements and the most convenient locations for their application will be also proposed. The drawn conclusions will contribute to the future development of rules concerning the design and implementation of structures that fulfil the idea of the textronic UHF RFID transponder and the collection of tools that are useful in this process.

Other research directions include the following: determining the impact of environmental factors on the transponder's operation arising from its proximity to the human body such as temperature or humidity; mechanical durability against interactions occurring during clothing usage, cleaning, and maintenance; or effectiveness during moments when such interactions occur, e.g., garment bending. The aforementioned studies would allow for a determination of the usefulness and applicability of the textronic UHF RFID transponder for RFID systems accompanying the user in everyday situations after the product purchase.

5. Conclusions

This article discusses the impact of the following factors on the textronic UHF RFID transponder's operation: system geometry, mutual inductance in coupling systems, displacement between the coupling circuits, and the resistance of the embroidered antenna. Displacements, both in vertical and horizontal directions, strongly affect the impedance of the transponder antenna and the transponder read range. A greater sensitivity to coupling circuit displacement in the vertical direction than in the horizontal direction was observed. This impact is presumed to be weaker for the dipole antenna, which will be verified in future studies. A small or no influence of the embroidered antenna resistance on the transponder antenna impedance and the transponder read range was detected, what means that above a certain value of the thread resistivity, no deterioration in the transponder's performance parameters is observed.

Author Contributions: Conceptualization, A.Z. and P.J.-M.; methodology, A.Z. and P.J.-M.; software, A.Z.; validation, A.Z.; formal analysis, A.Z.; investigation, A.Z. and P.J.-M.; resources, M.W. and P.P.; data curation, A.Z.; writing—original draft preparation, A.Z.; writing—review and editing, A.Z.; visualization, A.Z.; supervision, P.J.-M.; project administration, P.J.-M. All authors have read and agreed to the published version of the manuscript.

Funding: This research received no external funding.

Institutional Review Board Statement: Not applicable.

Informed Consent Statement: Not applicable.

Data Availability Statement: All the calculated and measured data will be provided upon request to the correspondent authors via email with an appropriate justification.

Conflicts of Interest: The authors declare no conflict of interest.

References

1. Wagih, M.; Balocchi, L.; Benassi, F.; Carvalho, N.B.; Chiao, J.-C.; Correia, R.; Costanzo, A.; Cui, Y.; Georgiadou, D.; Gouveia, C.; et al. Microwave-Enabled Wearables: Underpinning Technologies, Integration Platforms, and Next-Generation Roadmap. *IEEE J. Microwaves* **2023**, *3*, 193–226. [CrossRef]
2. Bouhassoune, I.; Chaibi, H.; Chehri, A.; Saadane, R. A Review of RFID-based Internet of Things in the Healthcare Area, the New Horizon of RFID. *Procedia Comput. Sci.* **2022**, *207*, 4151–4160. [CrossRef]
3. Ozek, E.A.; Tanyeli, S.; Yapici, M.K. Flexible Graphene Textile Temperature Sensing RFID Coils Based on Spray Printing. *IEEE Sensors J.* **2021**, *21*, 26382–26388. [CrossRef]
4. Yang, C.; Wang, X.; Mao, S. RFID-based 3D human pose tracking: A subject generalization approach. *Digit. Commun. Networks* **2022**, *8*, 278–288. [CrossRef]
5. Koski, K.; Lohan, E.S.; Sydänheimo, L.; Ukkonen, L.; Rahmat-Samii, Y. Electro-textile UHF RFID patch antennas for positioning and localization applications. In Proceedings of the 2014 IEEE RFID Technology and Applications Conference (RFID-TA), Tampere, Finland, 8–9 September 2014; pp. 246–250. [CrossRef]
6. Tao, X. *Wearable Electronics and Photonics*, 1st ed.; Woodhead Publishing Ltd.: Cambridge, UK, 2005.
7. Du, K.; Lin, R.; Yin, L.; Ho, J.S.; Wang, J.; Lim, C.T. Electronic textiles for energy, sensing, and communication. *iScience* **2022**, *25*, 104174. [CrossRef] [PubMed]
8. Arun, H. Advancements in the use of carbon nanotubes for antenna realization. *AEU Int. J. Electron. Commun.* **2021**, *136*, 153753. [CrossRef]
9. Tao, X. *Handbook of Smart Textiles*, 1st ed.; Springer: Singapore, 2015.
10. Zhang, Y.; Wang, H.; Lu, H.; Li, S.; Zhang, Y. Electronic fibers and textiles: Recent progress and perspective. *iScience* **2021**, *24*, 102716. [CrossRef]
11. Khan, A.; Haque, N.; Kabiraz, D.C.; Yeasin, A.; Al Rashid, H.; Sarker, A.C.; Hossain, G. A review on advanced nanocomposites materials based smart textile biosensor for healthcare monitoring from human sweat. *Sensors Actuators A Phys.* **2023**, *350*, 114093. [CrossRef]
12. Tekcin, M.; Paker, S.; Bahadir, S.K. UHF-RFID enabled wearable flexible printed sensor with antenna performance. *AEU Int. J. Electron. Commun.* **2022**, *157*, 154410. [CrossRef]
13. Dang, Q.H.; Nguyen-Trong, N.; Chen, S.J.; Fumeaux, C. Frequency-Reconfigurable UHF Wearable Textile Antenna for RFID Applications. In Proceedings of the 2023 5th Australian Microwave Symposium (AMS), Melbourne, Australia, 16–17 February 2023; pp. 23–24. [CrossRef]
14. Roggero, U.F.S.; Nista, S.V.G.; Hernández-Figueroa, H.E.; Mei, L.H.I.; Moshkalev, S.A. Graphene-biopolymer-based RFID tags: A performance comparison. *Mater. Today Commun.* **2022**, *31*, 103726. [CrossRef]
15. Zamil, N.; Ripin, N.; Ansari, M.; Jalil, Y.; Nordin, N.A.; Abdullah, I.; Yahya, Z.; Aziz, A.A. Design and Simulation of Dipole Patch Antenna using Graphene Sheet Substrate for UHF-RFID Application. In Proceedings of the 2021 IEEE Asia-Pacific Conference on Applied Electromagnetics (APACE), Penang, Malaysia, 20–22 December 2021; pp. 1–4. [CrossRef]

16. Chietera, F.P.; Colella, R.; Verma, A.; Ferraris, E.; Corcione, C.E.; Moraila-Martinez, C.L.; Gerardo, D.; Acid, Y.H.; Rivadeneyra, A.; Catarinucci, L. Laser-Induced Graphene, Fused Filament Fabrication, and Aerosol Jet Printing for Realizing Conductive Elements of UHF RFID Antennas. *IEEE J. Radio Freq. Identif.* **2022**, *6*, 601–609. [CrossRef]
17. Mostaccio, A.; Antonelli, G.; Occhiuzzi, C.; Martinelli, E.; Marrocco, G. Experimental characterization of Laser Induced Graphene (LIG) antennas for S-band wearable applications in 5G. In Proceedings of the 2022 IEEE 12th International Conference on RFID Technology and Applications (RFID-TA), Cagliari, Italy, 12–14 September 2022; pp. 51–54. [CrossRef]
18. Rivadeneyra, A.; Salmeron, J.F.; Rodriguez, N.; Morales, D.P.; Colella, R.; Chietera, F.P.; Catarinucci, L. Laser-Fabricated Antennas for RFID Applications. In Proceedings of the 2020 50th European Microwave Conference (EuMC), Utrecht, The Netherlands, 12–14 January 2021; pp. 812–815. [CrossRef]
19. Zhang, B.; Wang, Z.; Song, R.; Fu, H.; Zhao, X.; Zhang, C.; He, D.; Wu, Z.P. Passive UHF RFID tags made with graphene assembly film-based antennas. *Carbon* **2021**, *178*, 803–809. [CrossRef]
20. Zhang, B.; Zhang, C.; Wang, Y.; Wang, Z.; Liu, C.; He, D.; Wu, Z.P. Flexible Anti-Metal RFID Tag Antenna Based on High-Conductivity Graphene Assembly Film. *Sensors* **2021**, *21*, 1513. [CrossRef]
21. Song, R.; Chen, X.; Jiang, S.; Hu, Z.; He, D. Graphene Assembled Film Based Millimeter Wave Antenna Array for 5G Mobile Communications. In Proceedings of the 2021 IEEE MTT-S International Wireless Symposium (IWS), Nanjing, China, 23–26 May 2021; pp. 1–3. [CrossRef]
22. Gupta, D.; Sood, D.; Yu, M.; Kumar, M. Compact Biodegradable UHF RFID Tag for Short Life Cycle Applications. In Proceedings of the 2021 IEEE Indian Conference on Antennas and Propagation (InCAP), Jaipur, Rajasthan, India, 13–16 December 2021; pp. 399–401. [CrossRef]
23. Jaakkola, K.; Sandberg, H.; Lahti, M.; Ermolov, V. Near-Field UHF RFID Transponder with a Screen-Printed Graphene Antenna. *IEEE Trans. Components, Packag. Manuf. Technol.* **2019**, *9*, 616–623. [CrossRef]
24. Delipinar, T.; Ozek, E.A.; Kaya, C.E.; Tanyeli, S.; Yapici, M.K. Flexible Graphene Textile RFID Tags Based on Spray, Dispense and Contact Printing. In Proceedings of the 2020 IEEE International Conference on Flexible and Printable Sensors and Systems (FLEPS), Manchester, UK, 16–19 August 2020; pp. 1–4. [CrossRef]
25. Htwe, Y.Z.N.; Mariatti, M. Printed graphene and hybrid conductive inks for flexible, stretchable, and wearable electronics: Progress, opportunities, and challenges. *J. Sci. Adv. Mater. Devices* **2022**, *7*, 100435. [CrossRef]
26. Simegnaw, A.A.; Malengier, B.; Rotich, G.; Tadesse, M.G.; Van Langenhove, L. Review on the Integration of Microelectronics for E-Textile. *Materials* **2021**, *14*, 5113. [CrossRef] [PubMed]
27. Jiang, Y.; Leng, T.; Fang, Y.; Hu, Z.; Xu, L. Machine Embroidered Wearable e-textile Wideband UHF RFID Tag Antenna. In Proceedings of the 2019 IEEE International Symposium on Antennas and Propagation and USNC-URSI Radio Science Meeting, Atlanta, GA, USA, 7–12 July 2019; pp. 643–644. [CrossRef]
28. Liu, Y.; Yu, M.; Xia, B.; Wang, S.; Wang, M.; Chen, M.; Dai, S.; Wang, T.; Ye, T.T. E-Textile Battery-Less Displacement and Strain Sensor for Human Activities Tracking. *IEEE Internet Things J.* **2021**, *8*, 16486–16497. [CrossRef]
29. Yu, M.; Wang, S.; Liu, Y.; Xu, L.; Ye, T.T. Passive Embroidered Stretch Sensor Utilizing UHF RFID Antennas. In Proceedings of the 2019 IEEE SmartWorld, Ubiquitous Intelligence & Computing, Advanced & Trusted Computing, Scalable Computing & Communications, Cloud & Big Data Computing, Internet of People and Smart City Innovation (SmartWorld/SCALCOM/UIC/ATC/CBDCom/IOP/SCI), Leicester, UK, 19–23 August 2019; pp. 497–501. [CrossRef]
30. Khan, M.U.A.; Raad, R.; Foroughi, J.; Raheel, M.S.; Houshyar, S. An octagonal-shaped conductive HC12 & LIBERATOR-40 thread embroidered chipless RFID for general IoT applications. *Sensors Actuators A Phys.* **2021**, *318*, 112485. [CrossRef]
31. Khan, Z.; He, H.; Chen, X.; Ukkonen, L.; Virkki, J. Embroidered and e-textile Conductors Embedded inside 3D-printed Structures. In Proceedings of the 2019 Photonics & Electromagnetics Research Symposium-Fall (PIERS-Fall), Xiamen, China, 17–20 December 2019; pp. 1675–1680. [CrossRef]
32. Bakkali, M.E.; Martinez-Estrada, M.; Fernandez-Garcia, R.; Gil, I.; Mrabet, O.E. Effect of Bending on a Textile UHF-RFID Tag Antenna. In Proceedings of the 2020 14th European Conference on Antennas and Propagation (EuCAP), Copenhagen, Denmark, 15–20 March 2020; pp. 1–5. [CrossRef]
33. Pham, N.; Dao, N.C.; Chung, J.-Y. A text-meandered RFID tag implemented with conductive threads. *Microw. Opt. Technol. Lett.* **2016**, *58*, 1978–1984. [CrossRef]
34. Hanif, M.; Farhan, M.; Sharif, A. Design and analysis of flexible embroidered UHF-RFID tag on facemask for IoT applications using characteristics mode analysis. *AEU Int. J. Electron. Commun.* **2023**, *172*, 154940. [CrossRef]
35. Moradi, E.; Björninen, T.; Ukkonen, L.; Rahmat-Samii, Y. Characterization of embroidered dipole-type RFID tag antennas. In Proceedings of the 2012 IEEE International Conference on RFID-Technologies and Applications (RFID-TA), Nice, France, 5–7 November 2012; pp. 248–253. [CrossRef]
36. Kapetanakis, T.N.; Pavec, M.; Ioannidou, M.P.; Nikolopoulos, C.D.; Baklezos, A.T.; Soukup, R.; Vardiambasis, I.O. Embroidered Bow-Tie Wearable Antenna for the 868 and 915 MHz ISM Bands. *Electronics* **2021**, *10*, 1983. [CrossRef]
37. Patron, D.; Mongan, W.; Kurzweg, T.P.; Fontecchio, A.; Dion, G.; Anday, E.K.; Dandekar, K.R. On the Use of Knitted Antennas and Inductively Coupled RFID Tags for Wearable Applications. *IEEE Trans. Biomed. Circuits Syst.* **2016**, *10*, 1047–1057. [CrossRef]
38. Liu, Y.; Xu, L.; Li, Y.; Ye, T.T. Textile Based Embroidery-Friendly RFID Antenna Design Techniques. In Proceedings of the 2019 IEEE International Conference on RFID (RFID), Phoenix, AZ, USA, 2–4 April 2019; pp. 1–6. [CrossRef]

39. Liu, Y.; Yu, M.; Xu, L.; Li, Y.; Ye, T.T. Characterizations and Optimization Techniques of Embroidered RFID Antenna for Wearable Applications. *IEEE J. Radio Freq. Identif.* **2020**, *4*, 38–45. [CrossRef]
40. Benouakta, S.; Hutu, F.; Sette, D.; Duroc, Y. UHF RFID elastic textile yarn. *Microw. Opt. Technol. Lett.* **2020**, *62*, 3186–3194. [CrossRef]
41. Yang, Z.; Yin, C.; Wu, H.; Shimin, S.; Ba, L. Continuous Production of Conductive Fiber by Depressing Plateau-Rayleigh Instability for Wearable Smart Textile. In Proceedings of the 2021 IEEE 4th International Conference on Nanoscience and Technology (ICNST), Chengdu, China, 26–28 June 2021; pp. 1–4. [CrossRef]
42. Idumah, C.I. Design, fabrication, characterization and properties of metallic and conductive smart polymeric textiles for multifunctional applications. *Nano-Structures Nano-Objects* **2023**, *35*, 100982. [CrossRef]
43. Shekhawat, S.; Singh, S.; Singh, S.K. A review on bending analysis of polymer-based flexible patch antenna for IoT and wireless applications. *Mater. Today Proc.* **2022**, *66*, 3511–3516. [CrossRef]
44. Zahid, M.; Rathore, H.A.; Tayyab, H.; Rehan, Z.A.; Rashid, I.A.; Lodhi, M.; Zubair, U.; Shahid, I. Recent developments in textile based polymeric smart sensor for human health monitoring: A review. *Arab. J. Chem.* **2022**, *15*, 103480. [CrossRef]
45. Chen, X.; Ukkonen, L.; Bjorninen, T. Passive E-Textile UHF RFID-Based Wireless Strain Sensors with Integrated References. *IEEE Sensors J.* **2016**, *16*, 7835–7836. [CrossRef]
46. Lagha, F.; Beldi, S.; Latrach, L. Passive E-textile UHF RFID Tag for Wireless Body Centric Communications. In Proceedings of the 2018 30th International Conference on Microelectronics (ICM), Sousse, Tunisia, 16–19 December 2018; pp. 180–183. [CrossRef]
47. Chen, X.; Ukkonen, L.; Bjöninen, T.; Virkki, J. Comparison of E-textile dipole and folded dipole antennas for wearable passive UHF RFID tags. In Proceedings of the 2017 Progress in Electromagnetics Research Symposium-Fall (PIERS-FALL), Singapore, 19–22 November 2017; pp. 812–817. [CrossRef]
48. Ma, S.; Ukkonen, L.; Sydänheimo, L.; Björninen, T. Wearable E-textile split ring passive UHF RFID tag: Body-worn performance evaluation. In Proceedings of the 2017 IEEE Asia Pacific Microwave Conference (APMC), Kuala Lumpur, Malaysia, 13–16 November 2017; pp. 166–168. [CrossRef]
49. Thielens, A.; Baumbauer, C.; Anderson, M.G.; Ting, J.; Arias, A.C.; Rabaey, J.M. Feasability of On-Body Backscattering in the UHF-RFID Band using Screen-Printed Dipole Antennas. In Proceedings of the 2019 13th International Symposium on Medical Information and Communication Technology (ISMICT), Oslo, Norway, 8–10 May 2019; pp. 1–5. [CrossRef]
50. He, H.; Chen, X.; Mehmood, A.; Raivio, L.; Huttunen, H.; Raumonen, P.; Virkki, J. ClothFace: A Batteryless RFID-Based Textile Platform for Handwriting Recognition. *Sensors* **2020**, *20*, 4878. [CrossRef]
51. Wang, P.; Dong, L.; Wang, H.; Li, G.; Di, Y.; Xie, X.; Huang, D. Passive Wireless Dual-Tag UHF RFID Sensor System for Surface Crack Monitoring. *Sensors* **2021**, *21*, 882. [CrossRef]
52. Abdulghafor, R.; Turaev, S.; Almohamedh, H.; Alabdan, R.; Almutairi, B.; Almutairi, A.; Almotairi, S. Recent Advances in Passive UHF-RFID Tag Antenna Design for Improved Read Range in Product Packaging Applications: A Comprehensive Review. *IEEE Access* **2021**, *9*, 63611–63635. [CrossRef]
53. Luo, C.; Gil, I.; Fernández-García, R. Experimental comparison of three electro-textile interfaces for textile UHF-RFID tags on clothes. *AEU Int. J. Electron. Commun.* **2022**, *146*, 154137. [CrossRef]
54. Virkki, J.; Wei, Z.; Liu, A.; Ukkonen, L.; Björninen, T. Wearable Passive E-Textile UHF RFID Tag Based on a Slotted Patch Antenna with Sewn Ground and Microchip Interconnections. *Int. J. Antennas Propag.* **2017**, *2017*, 3476017. [CrossRef]
55. Jankowski Mihułowicz, P.; Węglarski, M.; Chamera, M.; Pyt, P. Textronic UHF RFID Transponder. *Sensors* **2021**, *21*, 1093. [CrossRef] [PubMed]
56. Jankowski-Mihułowicz, P.; Węglarski, M.; Wilczkiewicz, B.; Chamera, M.; Laskowski, G. The Influence of Textile Substrates on the Performance of Textronic RFID Transponders. *Materials* **2022**, *15*, 7060. [CrossRef] [PubMed]
57. Nizioł, M.; Jankowski-Mihułowicz, P.; Węglarski, M. The Influence of the Washing Process on the Impedance of Textronic Radio Frequency Identification Transponder Antennas. *Materials* **2023**, *16*, 4639. [CrossRef]
58. Griffiths, D.J. *Introduction to Electrodynamics*, 4th ed.; Pearson: London, UK, 2013; p. 246.
59. Piątek, Z.; Jabłoński, P. *Podstawy Teorii pola Elektromagnetycznego*, 2nd ed.; WNT: Warsaw, Poland, 2023.
60. Finkenzeller, K. *RFID Handbook*, 3rd ed.; Wiley: Chichester, UK, 2010; pp. 68–70.

MDPI AG
Grosspeteranlage 5
4052 Basel
Switzerland
Tel.: +41 61 683 77 34

Sensors Editorial Office
E-mail: sensors@mdpi.com
www.mdpi.com/journal/sensors